AF839158

ADVANCES IN CAD/CAM WORKSTATIONS

OTHER BOOKS BY THE EDITOR:

Advances in Engineering Data Handling: Case Studies, (Editor) Peter C. Wang, Kluwer Academic Publishers, Boston, 1984. ISBN 0-89838-154-1.

Advances in CAD/CAM: Case Studies, (Editor) Peter C. Wang, Kluwer Academic Publishers, Boston, 1984. ISBN 0-89838-157-6.

ADVANCES IN CAD/CAM WORKSTATIONS
Case Studies

Edited by

Peter C. C. Wang
Automation Technology Institute, Inc.
Pebble Beach, California

KLUWER ACADEMIC PUBLISHERS
Boston/Dordrecht/Lancaster

Distributors for North America:
Kluwer Academic Publishers
190 Old Derby Street
Hingham, Massachusetts 02043, USA

Distributors for the UK and Ireland:
Kluwer Academic Publishers
MTP Press Limited
Falcon House, Queen Square
Lancaster LA1 1RN, UNITED KINGDOM

Distributors for all other countries:
Kluwer Academic Publishers Group
Distribution Centre
Post Office Box 322
3300 AH Dordrecht, THE NETHERLANDS

Library of Congress Cataloging-in-Publication Data
Main entry under title:

Advances in CAD/CAM workstations.

Papers presented at the ATI Feb. 1985 symposium on automation technology held in Monterey, Calif.
Includes bibliographies.
1. CAD/CAM systems. I. Wang, Peter C. C.
TS155.6.A383 1986 670'.285 85-19848
ISBN-13: 978-1-4612-9403-0 e-ISBN-13: 978-1-4613-2273-3
DOI: 10.1007/978-1-4613-2273-3

CONTENTS

PREFACE vii

1. FLOCHART: A Solution to the Problem of Block Diagrams 1
   David A. Ackmann

2. Drawing Conversion: What? When? How? 11
   Thomas E. Bailey, Jr.

3. CAD Migration to the PC Environment — Tomorrow's Low Cost Workstation 25
   Robert T. Bannon

4. Advanced Systems Integration Management Via CAD Graphics 33
   Robert T. Bannon

5. Developing a Cartographic Geo-Code System (CAGES) 45
   Robert T. Bannon

6. CAD, CAM and GIS: Tools for Facilities Management and Planning 59
   William G. Brodie

7. Managing a Distributed CAD/CAM System 71
   Dale O. Christensen and Bryan L. Hill

8. The Impact of Geography and Distributed Data Collection on the Engineering of Army Telephone Systems World-Wide 79
   Coy Correll

9. CIMPLEX: Bridging the Gap Between "Computer Aided. . . ." and Computer Integrated Manufacturing 83
   Kenneth K. Dellenbach

10. The Impact of an Integrated Data Base on Automation in Mapping 95
    William L. Folchi

11. Update of Mass Storage Systems from the User's Point of View 107
    Jack K. Gingrich

12. Planning Precepts for CAD Instructional Design 117
    Cecil L. Groves

13. A Structural Model of Creative Process for Improved Interface Design 127
    Robin G. King

14. Navy CAD/CAM 145
    Captain John F. Leahy, III

15. Personal Workstations for CAD/CAM Applications 155
    Robert W. Lingard

16. Computer Graphics from the Graphic Design Perspective 165
    Aaron Marcus

17. The Digital Mapping Program of the U.S. Geological Survey 173
    Alan M. Mikuni

18. Automatic Scanning and Data Processing for CAD and Archival Systems 185
William Pferd, III

19. The Role of Human Factors in Introducing Graphic Artists to Computer Graphics 199
Betsy A. Riley

20. Graphic Standards: Status & Impact 209
Barry J. Shepherd

21. Graphic Related Standards 215
Barry J. Shepherd

22. Considerations for Establishing a Distributed Computer Graphics Mapping System 225
Larry D. Sieck

23. The Marriage Between CAD/CAM Systems and Robotics 229
Richard L. Simon

24. A Look at Technical Documentation Automation and Its Place in Distributed Data Base Information Systems 249
Richard C. Sisman

25. Successful Business Computer Graphics 265
Susan W. Ulicny

PREFACE

To understand what we know
and be aware of what is to be known
is a necessary approach to treating
CAD/CAM issues.

The challenge for all of us interested in CAD/CAM and engineering data handling is to understand what we know and what we need to know about today's and tomorrow's technology, to track the explosive development of our field and its broadening range of applications, to sort through the details which compete for our attention, and to perceive underlying trends.

A key development in the past year was the rapid and widespread acceptance by all user segments of personal computer-based CAD/CAM workstations, coupled with widespread use of software packages, both those developed for PC-based workstations and others converted from mainframe and mini systems for use on PC-based or 32-bit workstations. If this trend continues for a few more years, as much as 90% of all design work may be accomplished on advanced versions of PC-based workstations.

Many software systems vendors unknown until recently to the PC-based CAD/CAM community have now come to dominate the market—companies such as Autodesk, Chessell-Robocom, Future Net, T&W Systems, P-CAD, Cascade, 4-D Graphics, CADAM, Wang & Hornbuckle, and more than 20 other companies who sell PC-based CAD/CAM software.

The 25 papers in this book describe applications ranging from electrical, mechanical and architectural uses to solids modeling, finite element analysis and simulations. Most of the papers were presented at the ATI February 1985 Symposium on Automation Technology in Monterey, California. Similar symposia will be held in future years on topics related to general CAD/CAM and engineering data handling issues, and to other issues arising from applications of the technology, often in DoD environments—applications as seen in EDCARS, DSREDS and CAEDOS.

The symposia and the books in this series represent ATI's effort to bring forth some of the most important and critical issues in this field *from the user's point of view.* The symposia feature presentations by experienced professionals, and the books contain case studies of problems and solutions devised by practitioners. Because of the enormous potential impact on management and productivity advancement of CAD/CAM applications, documentation of the experiences of leading users, and coordinated reporting of these to others, can be of utmost importance.

The outstanding contributions presented here should help us understand what we know and help us become aware of what is to be known. I am deeply grateful to the authors of these papers, without whose dedicated work this volume would not have been possible. I hope this book will be of use in carrying their views to wider audiences, and in sharpening the thinking of us all about this remarkable and profoundly important field.

Peter Cheng-Chao Wang, Ph.D.
Editor

Automation Technology Institute
P.O. Box 242
Pebble Beach, CA 93953, U.S.A.

ADVANCES IN CAD/CAM WORKSTATIONS

1

FLOCHART: A SOLUTION TO THE PROBLEM OF BLOCK DIAGRAMS

David A. Ackmann
Monsanto Company
800 North Lindbergh Boulevard
St. Louis, Missouri 63167 U.S.A.

Abstract

A graphics program has been developed to aid in the creation of block diagrams. Such diagrams are useful as organization charts, process of information flow diagrams, pictographs, and in several other areas. The Monsanto system, "FLOCHART", allows the user to describe the desired output in a simple, flexible procedural language. The language allows shapes (polygons) to be placed anywhere on the page, and allows them to be of any color, size, or angle. Definitions for shapes ranging from a circle to an ear of corn are stored in libraries which can be enhanced by the user. Shapes can be connected easily. Text is also supported. FLOCHART is a FORTRAN program built around DISSPLA and is running under CMS and MVS. The system has proven to be very easy to learn, yet powerful enough for sophisticated usage.

1. INTRODUCTION

"Do you mean that you can't do a simple organization chart?" I had given the answer to that question about the graphics packages available at Monsanto many times. It seemed that block diagrams were in as much demand as bar charts or X/Y plots. I had tried to coerce TELLAGRAF with the "message" feature. I had experimented with its "text blocks". Nothing was really satisfactory. I did produce one good drawing with DISSPLA, but the cost of writing the FORTRAN program was not worth the final plot. SAS/GRAPH, GDM, and ICU were no help at all. I had to come up with a solution.

So I did. Monsanto now has a graphics package called "FLOCHART" which can create organization charts quickly and easily. It can also create pictographs. It has been used to create process and information flow diagrams. FLOCHART has even been applied in forms design and for general purpose cartoons (a cartoon is defined as a single shape intended to provide focus or emphasis in a presentation). Some samples of its output are shown in the first four figures.

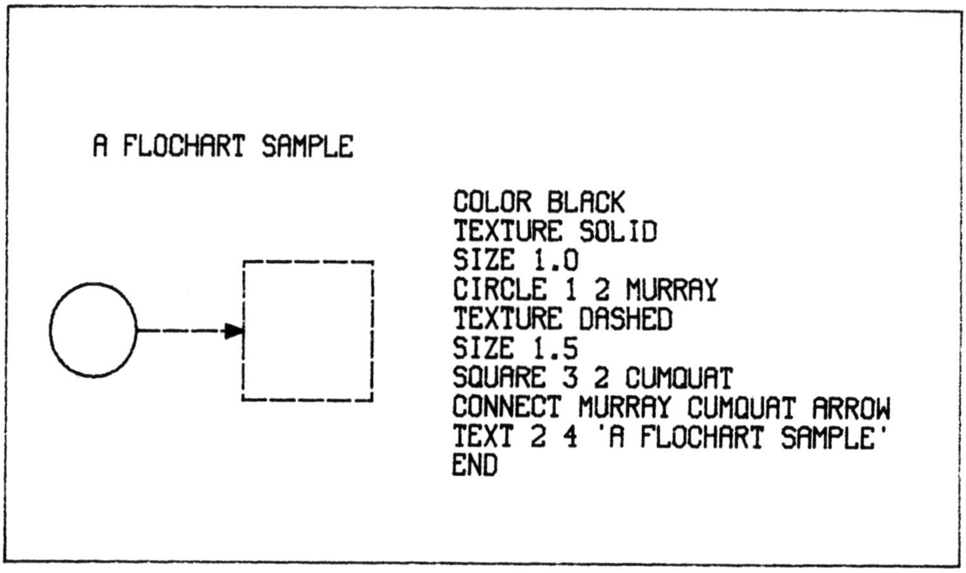

Figure 1. A FLOCHART Sample

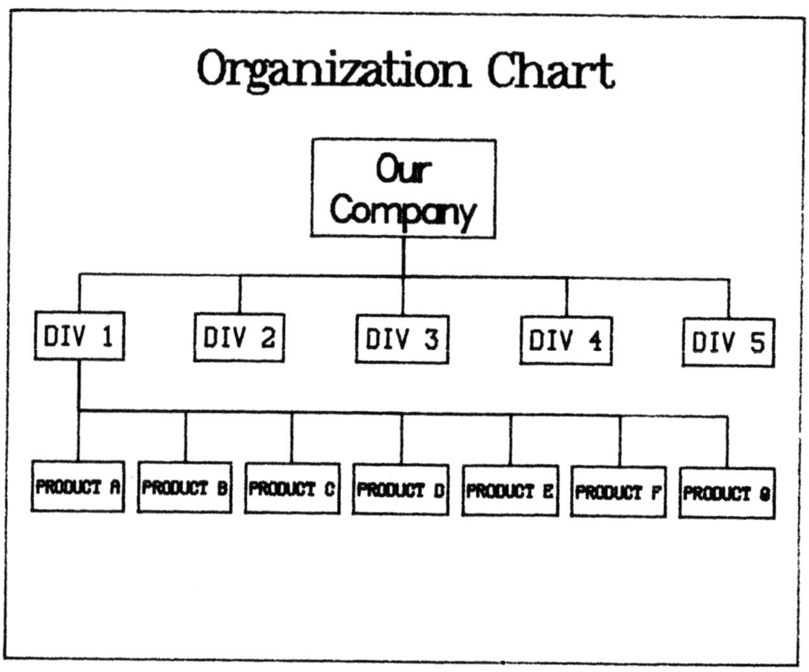

Figure 2. FLOCHART Produces Organization Charts

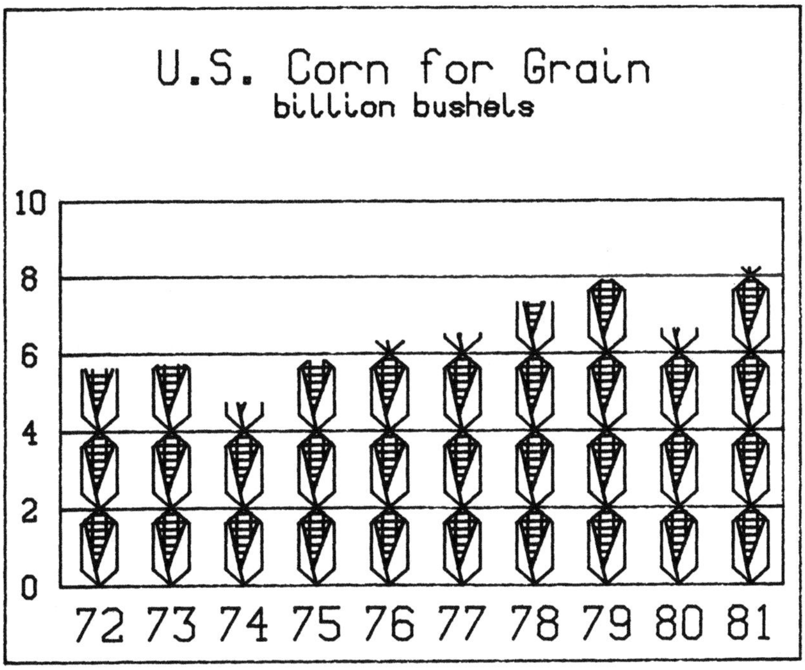

Figure 3. FLOCHART Produces Pictographs

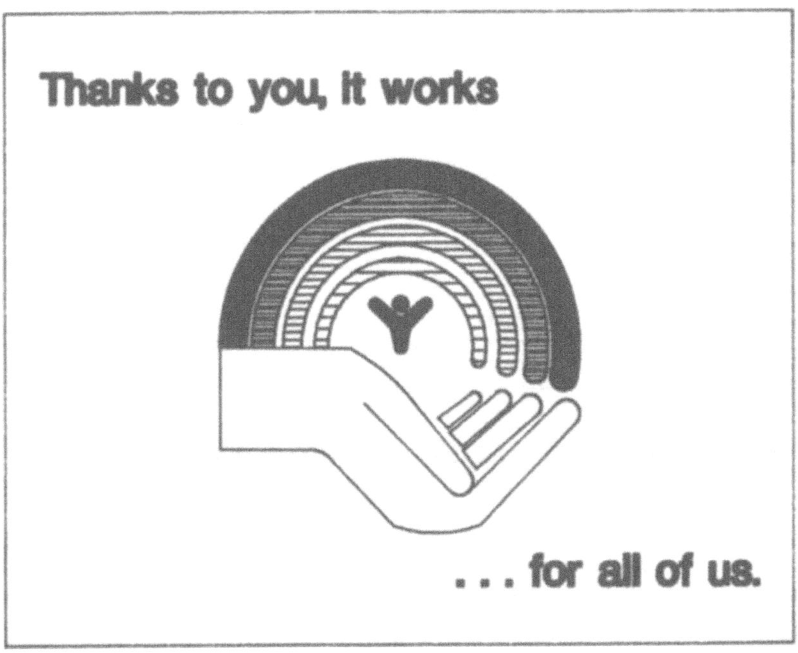

Figure 4. FLOCHART Produces Cartoons

Creating organization charts appears to be an easy task, but it is not. The task involves putting symbols (polygons) at any location on a page. Symbols often must be connected. They may take the form of simple shapes like circles or rectangles, but also may be much more exotic. The symbols must be able to change color, angle, or size. Text must also be suppported. It would be convenient to specify the location of a shape in units relative to the location of a previous shape, rather than in absolute coordinates. For a computer program to be able to create organization charts, it must be able to interpret the wishes of the user and be able to translate those wishes into instructions for a graphics device. To be accepted as a useful tool, it must be easy to learn, but flexible and powerful. To be user-friendly, it must be predictable. Finally, since I had no prior authorization to explore the idea, the program had to be very easy to develop. Since it was first conceived at Monsanto in February 1982, FLOCHART has met these requirements with great success.

2. IMPLEMENTATION

FLOCHART was implemented as a FORTRAN program built around DISSPLA. It interprets user commands, as entered via a command file, and processes them with appropriate calls to DISSPLA subroutines (a logical view of FLOCHART is shown below). Approximately 2000 executable lines of code are in the system.

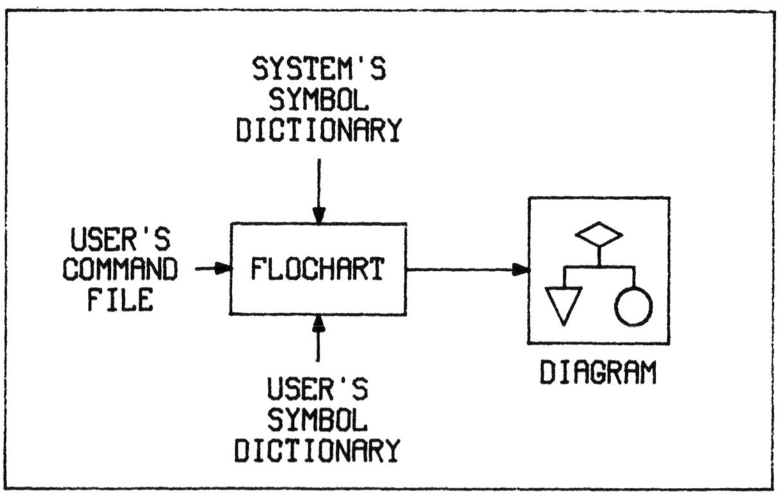

Figure 5. FLOCHART's Conceptual Design

The success of FLOCHART was the result of two design decisions. First, the user interface to the program would be via a new procedural command language specifically oriented toward block diagrams. And second, symbol definitions would be stored in data files and would be nominally sized within one square inch (1 inch = 2.54 cm). Let's examine these decisions. Exposure to systems such as TELLAGRAF caused me to think that a block diagram system should have a non-procedural command language. But this implied a sophisticated internal data base to keep track of all the directives until the signal is given to actually draw the plot. I was not willing to devote that amount of effort to the task, and so I did not immediately attack the project. After six months of deliberation, I came to the conclusion that there was nothing inherently wrong with a procedural command interface. This would allow me to process commands as they were encountered in the command file, thus eliminating the need to maintain anything but a very simple set of internal parameters.

The commands themselves were designed to be brief yet descriptive. To place a symbol on the page, all that was necessary was to specify the type of symbol desired and its X and Y coordinates (a name for the particular occurrence of the symbol was optional, but handy for connecting symbols). Other commands would change a symbol's size, color, or other attributes. Text commands would also exist. In Figure 1, the first four commands were responsible for the small, solid circle. In particular, the third line directed that a circle be drawn, centered 1.5 inches in and 5 inches up from the bottom of the page; the shape would be arbitrarily named "MURRAY". The next three commands drew the larger dashed square. The 8th line connected the two shapes (giving the symbols names made it easy to connect them). The 9th line drew the text, and the last line terminated the plot. With FLOCHART commands arriving in a procedural fashion, I could process them as I encountered them (when I found a "COLOR" command, I would handle it with a call to DISSPLA's "SETCLR" routine). This made the program much simpler to write than if the commands were non-procedural.

Next I had to consider the problems associated with symbol definition. There are at least three. How could I provide a set of standard symbols, yet still allow the user to add symbols of his own? How could I define symbols such that the user could predict exactly what the symbol would look like on the drawing? How could I control miscellaneous attributes of symbols such as the default color? These problems were solved through the symbol definition files.

First of all, every unique symbol in the system has a unique name (e.g., CIRCLE, DISK, or TRUCK). Next, observe that every symbol is nothing more than a series of points connected with straight line segments (yes, we will consider symbols with discontinous line segments). Finally, notice that some points in a symbol take on special significance; there may be a change in color at a point, or the point may be a logical place to connect this symbol with others. With these points in mind, let's consider what we must do to create a symbol dictionary.

Actually there will be two separate symbol dictionaries. One dictionary will contain general use symbols like circles and rectangles. This dictionary will be supported by the Central Computer Support Staff and will be available to all users. A second dictionary may be created by each user and will contain definitions for symbols that are peculiar to his job. In the symbol dictionaries there is one record for every point of every shape; each record contains the symbol's name and the X and Y coordinates for that point (it may contain additional information as well). The coordinates are defined such that the shape is either one inch long, or one inch high, or both. This means that its points will range from 0.0000 to 1.0000 in the horizontal and vertical directions. Furthermore, we will require that shapes be centered at the center of the square inch, at (0.5000,0.5000). A sample of the definition of a triangle is shown on the next page.

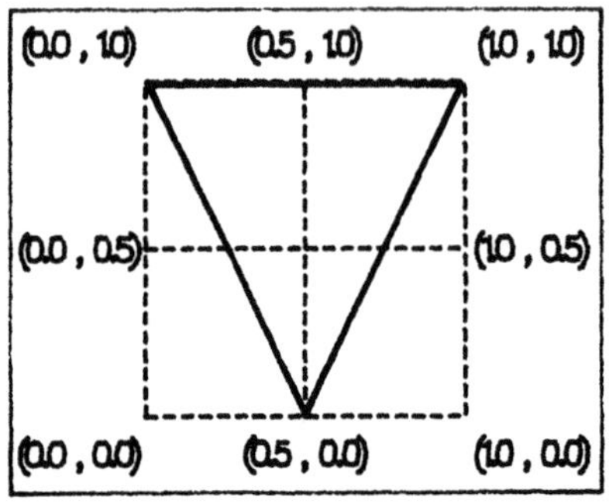

Figure 6. Definition of a TRIANGLE

For the sake of argument, assume that we want to create a horizontal drawing 11 x 8.5 inches in size. If we want to place a triangle centered at some point on the page (say 5 inches in and 2 inches up), we need only look at each point which is in the shape, subtract .5 from the X and Y coordinates of the point (remember that the shapes are centered at (.5,.5)), and then add back in the desired coordinates of the center. Try it. The top left point of a triangle is (0.0000,1.0000). Subtracting .5, we have (-0.5000,0.5000). Finally, adding in (5.000,2.000), we now know that the top left corner of the triangle should be placed at (4.5000,2.5000). After transforming all the points in the symbol in the same fashion, we then know the coordinates of every point required to draw the symbol in the proper position. Changing a symbol's size is now a simple matter of taking the coordinates after subtraction (-0.5000,0.5000) and multiplying them by the appropriate size before adding back in the desired center coordinates. For example, if the triangle were to have a size of 1.5, after multiplying (-0.5000,0.5000) by 1.5, we have (-0.7500,0.7500). Adding in the center gives us (4.2500,0.7500) as we expected. Other features of FLOCHART, such as rotating symbols or changing their aspect ratio, are an exercise in coordinate geometry.

There was a reason for my decision to center all shapes at (0.5000,0.5000) rather than (0.0000,0.0000). I established the convention that if there is a discontinuity in the definition of the shape, either the X or Y coordinate of the point which begins the new line segment must be negative. FLOCHART will use the absolute value of the coordinate as the location of the point, but will interpret its sign as an indication to move the pen rather than continue the current line.

One other part of symbol definition needs to be addressed. We need to have some way to handle all the "special cases" which always arise in any data processing problem. For example, we might want to give a shape a default color, or change its color right in the middle (I used this in defining the ear of corn). Let's decide that associated with each point in the symbol dictionary will be a few extra characters which may contain some special information. These characters may contain information to be acted upon when that point is encountered. In addition to color changes, I have used this feature to define certain points on a shape as connect points.

In summary, each point in every symbol in the symbol library exists as a record in the symbol library files. Each record contains the name of the symbol to which this point belongs, its X and Y coordinates, and a few extra characters which, if present, indicate some "special handling".

After defining the command language and deciding on how to define shapes, writing the actual program was fairly easy.

3. RESULTS

FLOCHART was subjected to test by a small group of graphics users for several months before it was released to the general Monsanto user community. The release was accomplished by making a presentation at an internal CMS users group meeting. A FLOCHART manual was made available at that time. In the first three months after the release, FLOCHART was executed over 4000 times and over 150 manuals were distributed. No classes have ever been taught because we have found that FLOCHART can be learned from the manual in less than 30 minutes. Often users have taught themselves how to use FLOCHART from looking at examples. In the first three months after its release, only one minor bug was encountered and that was corrected in a matter of a few minutes. The acceptance and ease of use of FLOCHART, coupled with its reliability, reinforces the initial design decisions.

Although it was originally intended for corporate organization charts, FLOCHART has been applied in many other areas. Logic flow diagrams in data processing areas were a natural application. Simple engineering diagrams (conceptual drawings only; FLOCHART is not a CADD system) came next. Pictographs became practical as the symbol library expanded beyond simple geometric shapes to more complex symbols. Cartoons began to appear. FLOCHART was next used to show the physical arrangement of equpment on the floor of the data center. Then it was used to show the logical paths between the various pieces of data processing hardware. Data processing used it to give a graphical representation to file layouts. FLOCHART was used in designing general business forms; it allowed us to eliminate several iterations between the group that needed the form and the printer. It has been used to produce Gantt charts. New users and new uses for FLOCHART are developing regularly. Some more examples follow.

4. CONCLUSIONS

I learned several things from the FLOCHART project. I found a need for a graphics system which would produce output beyond classical business/ presentation graphics. This need can be filled by systems which are driven by the user with a procedural language oriented toward the desired output. (Since FLOCHART, I have written four similar systems oriented toward some non-classical graphics which have applications in strategic planning.) New output formats will entice new users to computer graphics, users who had ignored it in the past due to a limited need for existing formats. Finally, I learned that it is not much harder to write a general purpose graphics package than one limited to a single specific application; all my future DISSPLA programs will be command driven instead of hard coded.

I expect some things to happen because of FLOCHART. I feel that it will cause us to begin to look at graphical input techniques in the business/

publication graphics area. This would allow the FLOCHART user to design a drawing with interactive rather than passive graphics. I expect pressure to create other systems oriented toward other graphical formats, particularly Gantt charts, each with their own peculiar command language. I expect the FLOCHART symbol library to grow in size and complexity. I expect FLOCHART to be "driven" by Data Base Management Systems; the DBMS would store information on the interrelationships of entities and pass to FLOCHART the directives to produce drawings of divisions or departments of interest.

At Monsanto, we now have a new tool which aids us in communication. We think tools like FLOCHART help us in our basic business functions.

5. BIOGRAPHY

Mr. Ackmann is a Senior Interactive Computing Consultant with Monsanto Company in St. Louis, Missouri. His responsibilities include training and support for users of analytical and presentation computer graphics on a large VM/CMS installation. He is a member of SIGGRAPH and NCGA and is a frequent speaker on computer graphics at various national conferences. Mr. Ackmann received a B.S. degree in Mechanical Engineering and an M.S. degree in Computer Science, both from the University of Missouri at Rolla.

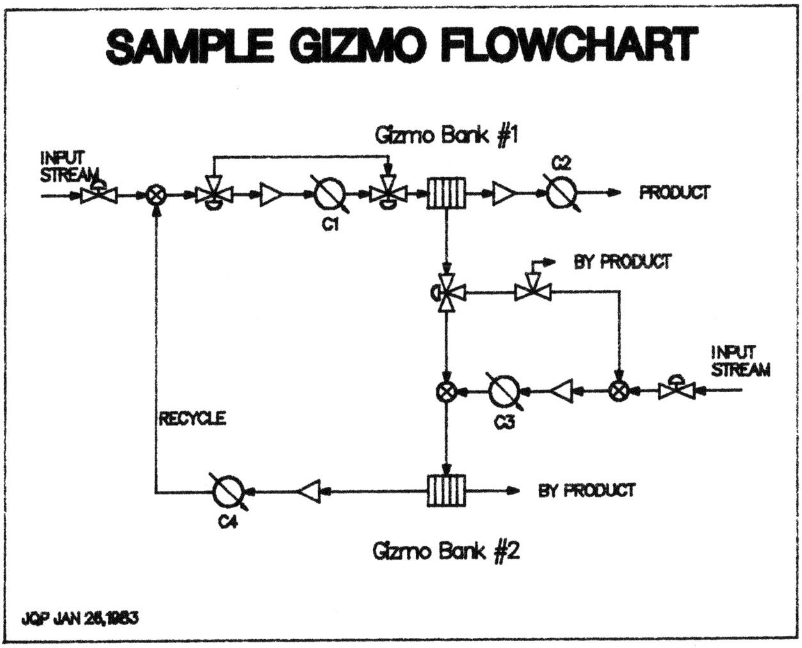

Figure 7: An Engineering Drawing

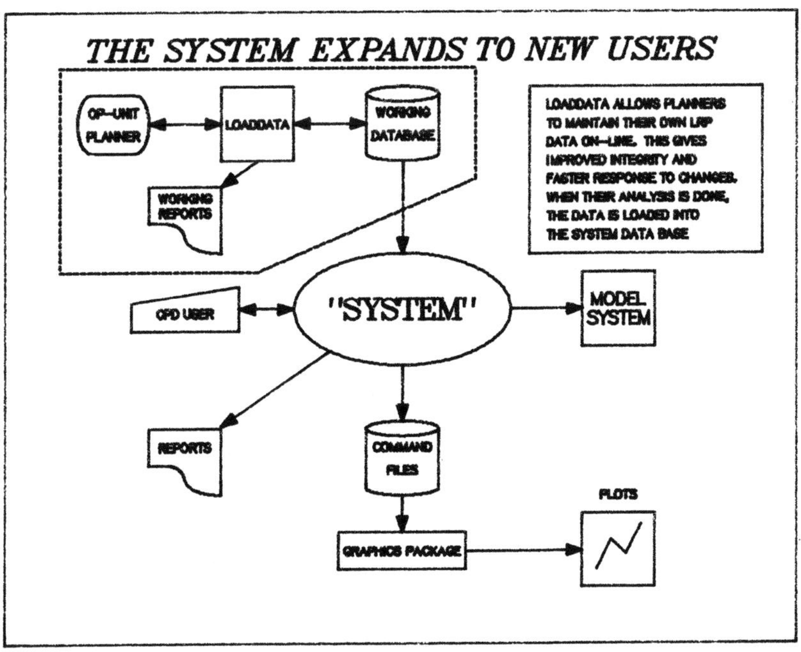

Figure 8: FLOCHART Shows How Data Flows In A System

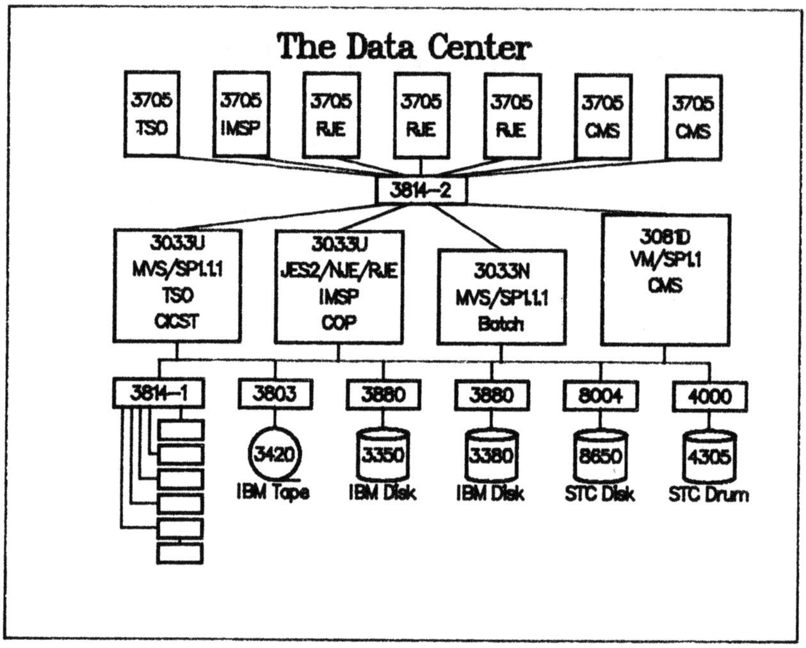

Figure 9: The Data Center

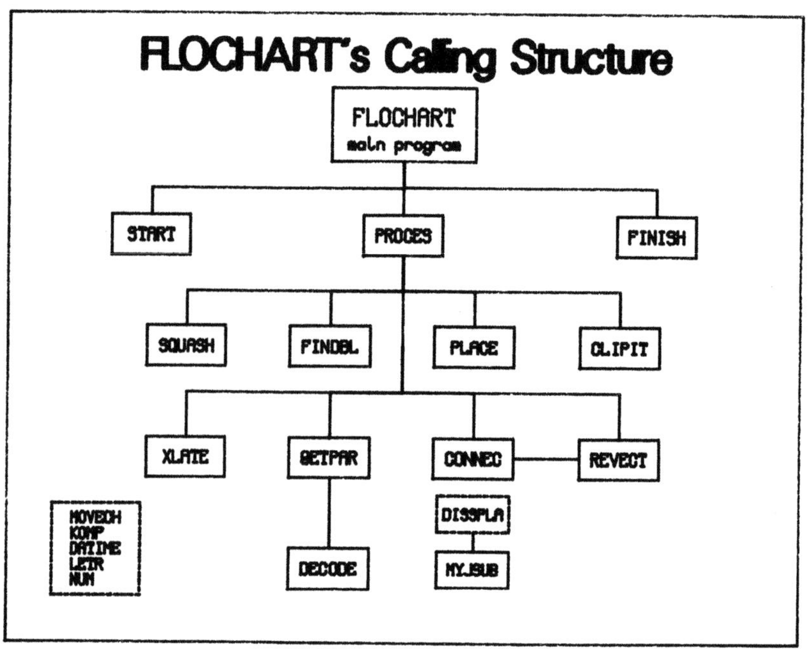

Figure 10: FLOCHART's Calling Structure

2

Drawing Conversion: What? When? How?

Thomas E. Bailey, Jr.
Metagraphics Inc.
30 Commerce Way
Woburn, Massachusetts 01801 U.S.A.

Abstract

Today many CAD/CAM users are beginning to enjoy unprecedented productivity improvements as a result of successfully automating their design and manufacturing processes. In order to maximize their productivity improvements and extend the benefits of CAD/CAM technology throughout their operations, these companies must establish comprehensive databases, a process that requires large-scale transfer of existing drawings to their CAD/CAM systems. The difficulty of establishing these databases has been a serious impediment to the extension of CAD/CAM benefits and to the effectiveness of CIM/CAD/CAM installations.

Recent developments are beginning to provide for effective transfer of drawing information at several levels. Contrary, however, to the popular concept that a single technological breakthrough would, without human intervention, allow for the automatic conversion of all paper drawings to CAD/CAM format, the most promising new approaches employ a variety of technologies and target separate, specific applications.

For most mechanical CAD/CAM applications, both accurate geometry and the drawing intelligence represented by dimensions and other notation on the drawing must be transferred to the CAD/CAM system. For these applications that require both database accuracy and intelligence, a new approach is required. This approach must combine a drawing scanner with software that includes:

(1) intelligent dimension and character recognition

(2) high performance editing capabilities

(3) advanced drafting features

(4) transfer mechanisms that provide compatible input for all major CAD/CAM systems.

While this paper's primary focus is on a new conversion solution for those demanding mechanical applications, it also highlighs systems that address other applications and levels of conversion.

Introduction

The tremendous growth of the CAD/CAM industry has made it the focus of attention from academia, to trade journals, to Wall Street. Since the 1960s when Ivan Sutherland first conceived of an interactive graphic system for drawing mechanical parts and the first storage tubes appeared, CAD/CAM growth has been phenomenal.

In the early 1970s the first turnkey CAD/CAM systems began to appear. CAD/CAM sales in 1984 total approximately $2.8 billion. By 1987, industry experts concur with a Yankee Group of Boston projection that the total CAD/CAM market will reach $6.9 billion. These figures represent a compound growth rate of over 45% sustained for the eight years starting in 1979.

CAD/CAM technology is spreading constantly to new industrial applications and new areas of the design and manufacturing process. Large CAD/CAM suppliers are diversifying their product offerings, and new product requirements offer specialized CAD/CAM vendors an increasing range of opportunities.

CIM/CAD/CAM. The ultimate result of all these trends has been called the Automated Factory, more recently referred to as CIM (Computer Integrated Manufacturing). An analyst with Dataquest Inc. of San Jose, California calls it the "big picture solution." The goal of such a facility is to use all of a corporation's resources in an integrated fashion to improve the productivity of an organization in designing and manufacturing higher quality products. The ultimate results of the Automated Factory will be better quality, shorter lead times, increased profits, and an improved return on investment.

Industry experts still have difficulty agreeing on the exact details of the Automated Factory. There is, however, agreement that at the heart of this new method of design and manufacturing is an integrated electronic database. This integrated database includes many types of product information to be used by all those needing the data within the organization, and, optionally, by vendors and subcontractors as well. It is the basis for design and manufacturing automation throughout the organization.

When today's potential customers consider purchasing CAD/CAM products, they immediately recognize the problem of converting their existing paper designs into the electronic databases they will need. These comprehensive databases of existing designs are important for three reasons: designs are used and modified for many years after the initial design is created; design in general is an evolutionary process with "new" designs being improvements or extensions of existing designs; and nearly all products include many components or subassemblies supplied by outside vendors. The Arthur D. Little consulting group did a study in 1983 which found that 80% of new designs utilize existing designs. A full 50% of these use existing designs directly.

CAD/CAM users have found that the problems of maintaining comprehensive, accessible databases are magnified many times over when their corporations have two different databases, one electronic and one paper, to maintain. A designer beginning work on a given part immediately confronts the difficulty of determining in which of the two databases the part is stored. Engineering Change Orders pose an additional problem because it is generally quicker and easier for the designer or engineer to change a paper drawing than to obtain CAD/CAM terminal time to make the change in the electronic database. So, even if a part was originally created on a CAD/CAM system, its database representation is obsolete if, as is often the case, only the paper drawing has been updated with the ECO. Moreover, subcontractor-supplied parts are not generally available in electronic format, and, even if tapes are available, they also are often not kept as up-to-date as the printed paper descriptions.

It is becoming clear that the dual database problem begins with the introduction of the first CAD/CAM system into a facility, and continues for many years after all ongoing design efforts are integrated in the CIM environment. The need for a method for integrating those two databases has become critical: CAD/CAM users must be able to access, from their terminals, all the information they need, much of which currently is available only on paper.

A solution for the dual database problem is to provide a method for readily converting all paper designs into the electronic database. An automated drawing conversion system that could accurately recreate mechanical design information would accelerate the transition to a single electronic database that is critical to CIM/CAD/CAM productivity.

Drawing Conversion

The ideal: A single, "magic" solution. The ideal solution to the paper drawing conversion problem was once described by Dr. Joel Orr, CAD/CAM industry consultant, as a "magic box" into which drawings would be fed. This device would read the drawing information and automatically create an accurate electronic database. Unfortunately, no one has been able to implement this solution for many reasons:

(1) The most important reason is that as the final product design is created, produced, and tested, it goes through several iterations in which actual dimensional notations are corrected, but the geometry on the drawing usually is not updated.

(2) Drawings are created on paper following industry standards in a "nearly" precise manner, but they may not contain all the required information.

(3) The physical condition of the paper drawings may cause reader input inaccuracies.

The "magic box" concept traditionally has been envisioned as a hardware device that would somehow eliminate the need for human intelligence and drawing interpretation. Today's high resolution scanners are, to a large extent, the result of this concept: they have attempted to so accurately recreate the lines on a drawing that the need for intelligence would become secondary. This approach has worked for some applications such as mapping and technical publications, but it does not meet the criteria for engineering applications.

To compensate for drawing inadequacies mentioned earlier and for other problems, two things must occur to accurately capture, in electronic format, the meaning of a mechanical drawing. First, special software must use the notational information on a drawing to correct the geometric information. Second, a person must be involved in the drawing conversion process to give direction, edit the drawing, and evaluate the results.

Emerging solutions: Levels of drawing conversion. Many companies, particularly those that are working toward the automated factory, are beginning to cite improved designs as their primary CAD/CAM benefit. These improved designs are occurring through the use of engineering analysis and simulation techniques that subject designs, before production, to a degree of precise scrutiny that would be impossible without a CAD/CAM system. This increasingly demanding use of CAD/CAM systems has a major impact on the approach to drawing conversion.

There is no "magic box" technology that can convert drawings for all applications. Most drawing conversion systems are still in the development stage, and, since all of them include basic scanning and digitizing techniques, it is difficult for CAD/CAM users to differentiate among them. In order to evaluate the emerging techniques that may be used to transfer drawings into electronic format, CAD/CAM users must be able to focus on the systems that are designed to solve their specific conversion problems.

An effective way to segment drawing conversion techniques and to define the role of drawing conversion in the overall CAD/CAM effort is to analyze the separate levels of emerging drawing conversion systems according to the relative intelligence of the databases they produce. To that end, a method of categorizing the intelligence level of each database produced has been developed, as shown in Figure 1.

DRAWING CONVERSION

DATABASE LEVELS

INTELLIGENCE LEVEL	DATABASE	USEFULNESS
5	Full 3D Model	Future, Used with Solid Modeling Systems
4	Dimensions, Geometric Primitives, Symbols, Alpha Numerics	Mechanical Applications
3	Geometric Primitives, Symbols, Alpha Numerics	Schematic Representations (P&ID, PC, WD)
2	Vectors	Mapping Applications
1	Pixels	Graphics Arts, Tech Pubs, Archival Storage

metagraphics
Building CAD/CAM Databases

Figure 1

A Level 1 Drawing Conversion product scans a drawing to provide a pixel, or raster representation. There are billions of engineering drawings currently stored on paper or film, and many companies need only a compact archival storage medium that will allow them to display drawings, or possibly transmit them from location to location for review. A Level 1 product could fit this requirement by providing a basic scanning and archival storage system similar to a reprographics system in pure drafting environments. With the combination of a medium resolution scanner, a relatively simple database management system, and an optical disc storage unit of one to one hundred gigabytes, almost all of a company's drawings can be safely and compactly stored for later retrieval and/or transmission. This technology will begin to mature in two to three years and will complement products at levels three and four, which provide real CAD/CAM databases.

Level 2 products convert a pixel (raster) representation of the drawing into a series of vectors in an attempt to create a database which can be read by a CAD system. Level 2 probably has the least potential usefulness to CIM/CAD/CAM users of the five levels defined here. Everything at this level is translated into vectors, so circles, arcs, dimensions, and all notations are represented by series of small vectors. As the resolution of the scanner increases with the accuracy required by the user, so does the number of vectors required to generate the pixel image. The resulting database is very large and difficult to manage, and its intelligence is minimal. Only in certain mapping applications and some technical publication applications is this database useful.

A Level 3 database is one in which a level of intelligence begins to appear. Instead of relying on pure vectorization, Level 3 conversion systems intelligently represent geometric entities such as circles and arcs. Viable systems at this level should be able to transmit alphanumeric information, but there is no provision at this level for an intelligent relationship between the notation and the geometry. The accuracy of a Level 3 database is absolutely dependent upon the accuracy of the geometry on the drawing. Potentially, these systems can take in incomplete or inaccurate drawings and reproduce them as inaccurate or incomplete electronic databases. These files, however, can be read by a CAD system and then edited. For installations that use CAD systems exclusively, or primarily, as drafting tools, a Level 3 database may be useful.

In Level 4 databases the intelligence level is increased and accuracy is added. Geometric elements and alphanumeric information, such as dimensions in the case of mechanical drawings, are combined with the descriptive, geometric information on the drawing. This degree of intelligence provides a unique level of database accuracy, particularly for mechanical applications and other users who are taking advantage of the advanced engineering and analysis features of their CAD/CAM systems. A Level 4 system reproduces the drawing geometry. At Level 4, however, a database representation of a mechanical drawing is no longer limited by the accuracy of the drawing geometry, but instead becomes as accurate as the dimensions and other notation on the drawing. If the intent is to produce a CAD "model" in order to do finite element analysis, NC programming, or circuit simulation, a Level 4 database is required. Metagraphics Inc. supplies a product that produces a Level 4 database.

A Level 5 database is one that a drawing conversion system might produce by taking two dimensional drawings and creating an accurate solid model. As the use of solid modeling systems increases, the need for an easy method of converting from 2D to 3D design operations will expand. A major drawback of most current solid modeling packages is in the user interface which requires designers and drafters to think and operate in three dimensional space, using boolean algebra just as the computer does. If a front-end conversion system could allow user-entry in traditional 2D terms and then, using an automated process, convert the information into a solid model,it might help enlarge the market for solid modeling systems.

It is very important for potential drawing conversion users to consider how the data produced by a drawing conversion system is going to be used on their target CAD system. Most CAD systems are justified on their modeling capabilities, not on their ability to do drafting. Companies look forward to automating such tasks as NC, finite element analysis, robotics, mass property calculations, interference studies, and circuit simulation. These applications, along with the ability to automatically extract accurate information from the model, produce the CAD/CAM system payback, so it is reasonable to predicate system productivity on their effectiveness.

To take full advantage of these advanced CAD/CAM features, CAD/CAM users need drawing conversion systems that produce comprehensive, accurate databases that can support advanced CAD/CAM engineering and modeling techniques. These are Level 4 or Level 5 systems. There are no Level 5 systems today.

Keys to effective drawing conversion systems. A useful conversion system must create, cost effectively and accurately, intelligent database input in formats that are acceptable to the CAD systems it serves. The keys to creating these accurate, intelligent databases efficiently are similar to those required of an effective CAD/CAM system. The system must be so friendly that it is immediately productive -- allowing users to get on with the job of creating databases without a lengthy training period. And the system must, itself, incorporate genuine intelligence so that minimal operator input will be required to interpret and convert even complex drawings.

In the case of drawing conversion, accuracy is derived from a combination of effective hardware and software that is designed for the specific job at hand. A system programmed to use dimensional data to verify geometry, for example, can provide much more accurate drawing files than one that deals only with the geometry. In applications, for which limited accuracy is sufficient, a Level 3 product may be sufficient. For most mechanical applications, however, the greatest possible degree of database accuracy and intelligence is needed.

Because the databases must be established if CAD/CAM systems are to provide the productivity they promise, cost-effectiveness for drawing conversion systems is best measured against the cost of creating the necessary databases using a CAD system. Moreover, particularly for Level 4 purposes, the degree of accuracy and comprehensiveness required can only be supplied by a CAD system or the most advanced drawing conversion system. The relative cost-effectiveness of the drawing conversion system is derived from the factors that make it both accurate and productive. Primary among these are knowledge engineering and user friendliness.

Making systems user-friendly will be a primary determining factor in the growth rates of CAD/CAM systems and in the growth of the drawing conversion segment of the industry. Today's CAD/CAM users want systems that are easy to operate and that require little or no training. These CAD/CAM operators see their systems in very practical terms -- as necessary tools to be purchased only if they can be immediately useful. As Business Week reported in its March 27, 1984 issue:

> "Today, making a program simple is just as important as what the computer actually does." Even mainframe vendors, the article continued, "..have been forced to make software easier, because users have been spoiled by micros."

Progress is being made, but most CAD/CAM systems today are not so easy to learn or operate as they need to be. Most require lengthy, specialized training. This training factor is closely tied to system productivity. Experts point to the scarcity of trained operators as a serious impediment to purchase for many CAD/CAM prospects who know that the systems are not yet friendly enough to allow early productivity after rapid training of their people.

While the current approach to the problem of difficult-to-operate CAD/CAM systems is to allow for extensive operator training, the emerging solution is to provide systems that are inherently easy to learn and easy to use. These "friendly" systems will provide not only more immediate productivity, but also they will encourage existing users to take full advantage of their systems. Once system fundamentals are mastered, designers can begin to use the additional application packages and advanced CAD/CAM features that add new design and analysis capabilities. With these overriding considerations present at nearly all major CAD/CAM installations, the challenge is critical for drawing conversion suppliers: their systems must be so easy to use that while offloading the task of database creation from the CAD system they put no demand on a company for the scarce resource of trained CAD operators.

What is a "friendly" system? In the CAD/CAM arena, it is a system that is designed for and adapted to a specific engineering application and the workstyle of operators in that application. This must be true both from a physical, ergonomic, point of view and in the way the computer and the interface operate.

What are the elements of a friendly system? One aspect is the traditional ergonomic assessment of the physical equipment. It must be designed for the work that will be done with it. If it is a drafting or a drawing conversion system, the work surface must accommodate sketches and full size drawings. The placement of the screen, the mouse or pen, and even the chair may affect the productivity of the user, and these must be easy for the system operator to modify.

The most important part of the system is the operator interface. Computer performance is continually improving and specialized software programs that perform an increasing number of design, analysis, and manufacturing and drawing conversion tasks are proliferating. To maximize the effectiveness of these developments, each interface must be designed to eliminate "computerese" and explain clearly, in language appropriate to the application, what the computer is doing. The user must be allowed to control the system easily. As an article in the September/October 1983 issue of Computer Graphics News put it:

> Many computer systems, CAD/CAM especially, are Rolls Royce machines with VW dashboards. The explanatory systems are simply not yet designed.

A great deal of work is going into creating the kind of interface that will make computer systems easier for operators, particularly for novice users.

A great many factors contribute to the friendliness and usability of computer systems. Computer response time is critical. Effective displays, particularly color displays, are also very important in general CAD/CAM type applications. And, truly friendly systems must meet the needs of experienced as well as novice users. To meet that goal, the full range of on-screen menus and documentation tutorials must appear only for users who need them. Experienced designers must be able to bypass some steps in a design process that may be absolutely necessary for a beginner.

Systems can be designed to offer an appropriate interface to both casual and expert users. This can be accomplished through the interactive creation of specific user preference files and through built-in intelligence that allows the system to respond as though it is making an assessment of the level of user expertise. If, for example, the average time between system response and the subsequent operator instruction exceeds a defined level, the system may be designed to assume that a novice or casual operator is in control and offer full menu and help options. If, on the other hand, an operator consistently moves to the next action in less than the defined period, the system may modify its behavior, offering less help and speeding up the process at hand.

Frequently the term "artificial intelligence" or "knowledge engineering" is applied to any computer activity in which the computer appears to be thinking like a human being (i.e., performs symbolic inference) rather than performing a calculation or retrieving data. In CAD/CAM applications, knowledge engineering has the potential to create better application-specific programs and workstations. In the same way that dedicated computers have increased workstation performance, dedicated task systems, via built-in expertise, can provide better, more effective solutions to a range of specified problems.

Knowledge engineering will allow the creation of truly "expert" or "goal seeking" systems. By incorporating both the facts necessary for a specific application and the heuristic techniques of reasoning for the discipline involved, a computer system can be created that assists its operator in thinking through and solving problems. To establish both the knowledge base and the inference procedure for an engineering discipline, a computer scientist needs to work closely with experts in that discipline to define and implement the required knowledge and logical procedures. Knowledge engineering is time-consuming and difficult, but the promise of systems that can help even novice designers and engineers make use of built-in expertise is worth the effort.

The Metagraphics approach: A realistic solution for engineering applications. The need for a drawing conversion system that incorporates these new developments is clear. The technology now exists to provide it. The CAD/CAM industry has explored the limits of hardware approaches that produced inadequate digitizers or scanners that could simply photograph drawings, but not convey the intelligence they represent. There is no hardware solution, no "magic box," but the objectives of that ideal solution can be accomplished by a user-friendly system that combines hardware, traditional CAD software approaches, and both artificial and human intelligence in order to transfer both the image of a drawing and its meaning as well.

Taking advantage of several technologies, Metagraphics has built a system that scans a drawing for quick image input and then applies a number of new software techniques to create an intelligent CAD/CAM database representation of the part. The process is fast and cost-effective.

Metagraphics has designed its Level 4 system specifically for CAD/CAM engineering drawing conversion, profiling expected users to make it responsive to their needs. Its friendly user interface makes it easy to learn and operate.

By designing a friendly system that is truly flexible, Metagraphics has targeted the needs of an almost limitless range of novice to frequent users. By incorporating real intelligence into the M-4200 system, Metagraphics has effectively targeted the specific task of drawing conversion. Knowledge engineering provides unique levels of assistance in three significant areas: character recognition, text-context recognition, and Geometric Rectification.

The field of intelligent character recognition and text-context recognition is receiving attention as a software frontier. Like other areas that employ new techniques in artificial intelligence, intelligent character and text-context recognition is beginning to offer new capabilities, speed, and efficiency to a range of applications.

The Metagraphics system optically recognizes hand printed alphanumeric characters as well as ANSI standard CAD-generated notation. The system is engineered to understand the meaning and context of dimensions just as a drafter would. With this built-in intelligence the Metagraphics-4200 can very rapidly reproduce a meaningful part representation with true associativity between geometry and notation. Unlike other conversion methods that turn images of characters into series of vectors to reproduce shapes which must, in turn, be decoded by the CAD system, the Metagraphics system understands drawing notation and uses it to create the CAD/CAM part file. An additional benefit of this intelligent character recognition is that Metagraphics part files are transmitted to CAD/CAM systems as standard size files, not the prohibitively large vector files common to scanning.

The M-4200 is an intelligent drawing conversion system that effectively assumes responsibility for interpreting a scanned part and representing in a CAD/CAM database format. In most engineering applications, such representations must be absolutely accurate, based on dimensions and notation, because generalized geometric data or simple part images are of extremely limited use.

The process the M-4200 uses to construct dimensionally accurate, unambiguous CAD/CAM part drawings from scan data is called Geometric Rectification. Because the M-4200 system is built with a knowledge of drafting procedures, it can intelligently make accurate assumptions regarding the meaning of a drawing, how the geometry actually is constructed, and how views are related to one another. The system interprets a drawing as an experienced drafter would, so the inexperienced operators can create CAD/CAM database drawings many times faster than even practiced operators of CAD/CAM workstations.

In most cases, once the edit station operator instructs the system to accept the dimensions on the scanned drawing, the system rectifies the drawing without further assistance. If, the original drawing is ambiguous or contains errors, however, the system will ask the operator to help resolve these problems. In these cases, the system will ask for additional information and/or for confirmation of assumptions it is making until it has eliminated all possible design interpretations except the correct one.

Using the M-4200, CAD/CAM users can create databases whose quality can be equalled only by using a CAD/CAM terminal to redraft existing parts. By off-loading the enormous task of drawing conversion from the CAD/CAM system, however, CAD users get maximum productivity from their design systems as well as faster, more cost-effective database construction.

A typical Metagraphics-4200 installation consists of one scan station networked to a number of edit stations and cabled to peripherals, including plotters and tape drivers (See Figure 2). The conversion process is separated into three stages of operation: scanning, editing, transferring data.

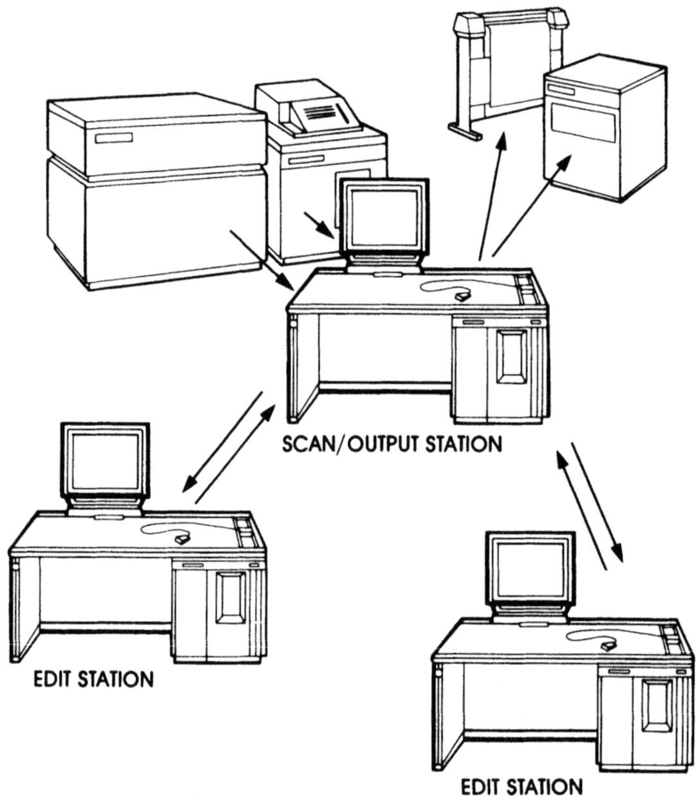

Figure 2

Using a flatbed paper scanner or aperture card reader, the M-4200 scans a drawing which then is reviewed at a high-resolution graphics terminal. Sophisticated software performs intelligent character recognition that interprets the notational data on the drawing, including all dimensional information. The drawing image and its notational intelligence are then transferred to an edit station.

The Edit Station is a high-performance, high-resolution color graphics workstation. An operator at this station verifies the drawing information and instructs the computer to create a CAD/CAM file. Because the system is application-oriented and understands the meaning of notational data and drawing geometry, the edit station operator does not need to laboriously recreate the part or instruct the system regarding its meaning. As a result, the M-4200 conversion process is extremely fast and easy to complete.

If it is necessary to change the drawing by adding new information, deleting data, or revising its format, the system guides the operator through the steps that will create exactly the drawing that is needed. In mechanical applications where engineering changes to dimensions are commonly inserted, but geometry is rarely redrawn to reflect them, the M-4200 offers a unique opportunity to improve the accuracy of the drawing before sending it to the CAD/CAM system. Once the operator verifies the dimensional information and modifies or deletes inaccurate or redundant dimensions, the system recreates the geometry precisely to the dimensional specifications.

In this new era of increasingly successful CAE/CAD/CAM, drawing conversion has become a point of focus almost overnight. Unknown as recently as two years ago except as an ideal "magic box," drawing conversion systems today are poised to provide a surge of productivity improvements for CAD/CAM users in a variety of industries. By directing new technical developments at a clearly defined need it has become possible to create a system that does what recently appeared to be impossible -- translate, accurately and cost-effectively, the meaning of engineering drawings to CAD/CAM system databases. Continuing developments in fields such as knowledge engineering and system interface design are enlarging the range of applications to which this new technology can be applied. Without drawing conversion, Computer Integrated Manufacturing draws automated design and manufacturing together. With drawing conversion, CIM can increase its scope as well as its efficiency by integrating the expertise of all design and engineering areas, including those that are not yet automated.

BIOGRAPHY

Thomas E. Bailey, Jr. is president of Metagraphics Inc., a company that supplies automated drawing conversion systems for the CAD/CAM industry. Before founding Metagraphics in 1982, Mr. Bailey led the development and marketing of Computervision CAD/CAM systems for mechanical design and manufacturing applications, first as Director of Application Software, then as Director of Product Marketing, and later as Vice President of Mechanical Systems Business. During his tenure at Computervision (1975-1982), the company captured a greater than sixty per cent share of the mechanical CAD/CAM market.

From 1971 to 1975 Mr. Bailey helped develop CAD/CAM systems at Gerber Scientific. His engineering experience includes engineering simulation and test at Pratt and Whitney and Lockheed. Mr. Bailey has a BSEE from Virginia Tech., an MS in Computer Science from RPI, and an MBA from Northeastern University.

3

CAD Migration to the PC Environment - Tomorrow's Low Cost Workstation

by
Robert T. Bannon
Manager- Systems Engineering - Computer Graphics Planning
AT&T Communications
Bedminster, N.J., USA

Abstract

The purpose of this paper is to examine the downward migration of Computer Aided Design (CAD) applications from the mainframe and mini-computer environment to the 32-bit micro-based processors.

The price performance of the low cost CAD workstations and the personal computers offer an economical alternative to some of the more expensive turnkey CAD systems. Depending on the application, user functional requirements and the display output resolution requirements, the CAD microprocessors have achieved a degree of sophistication, and sufficient productivity increases are attainable to justify the mechanization of various design functions that previously would not support a mini-computer based CAD solution.

Introduction

Many companies are being faced with new "profit center concepts" where productivity and cost reductions have become the keywords. Since labor related costs are a primary contributor to high budgets and cost overruns, a solution to cost control and productivity increases is mechanization of the process. This is especially true in a design environment, whether it supports mechanical, electrical or architectural layouts.

Large corporations have the work volume and the capital resources required to support large scale CAD systems, which range in cost from $300,000 to over $1 million. These large CAD systems often require a conditioned environment which adds additional cost to the CAD introduction. However, smaller companies do not have the same funding capabilities, but they are faced with the same productivity increase issues.

Many CAD application opportunities have been ignored or turned down in the past even though productivity increases ranged from a 3 or 4 to 1 improvement instead of being in the 10 to 1 improvement range that is achievable with many of the larger CAD systems. This mechanization rejection is in part due to the initial costs of introducing a CAD system. These high initial costs associated with a turnkey system reflect only the purchase of hardware and software and do not include training costs. With the majority of architectural firms and mechanical design shops employing less than 20 people, a $300,000 plus CAD is not an economical solution to the problem.

Recent hardware developments in microprocessor based systems - personal and business computers - and new software packages have now resulted in the emergence of low cost CAD workstations. The introduction of the new products are geared to meet small company or specific application needs. Software packages are available for the IBM PC/XT or the AT&T PC 6300, both of which are compatible computers, ranging from $495 to $10,000, and fully integrated CAD workstations are available for approximately $30,000 to $55,000.

Low Cost Applications

Many small architectural firms, the one- and two-man offices, as well as some of the larger firms, have bought PC's to meet their business graphics and spreadsheet requirements. However, until recently, purchasing a CAD system required the in-house development of a specific CAD application or the purchasing of an integrated package in the $300,000 plus range. This investment exceeded the "two-year payback rule" which is employed by many companies for computer system acquisition. Even though the expected productivity increase was in the range of 7-10 to 1, the volume of work would not support a large CAD system. Now these firms can utilize the same PCs on which they run their spreadsheets for CAD drawings.

Less demanding drawings such as those requiring up to approximately two dozen standard units are ideal applications for the PC or low-cost workstation approach. The computer can draw as many standard modules as instructed with exact precision. Regardless of whether the unit is in a "typical" position, mirrored or rotated, the microprocessor has the ability to draw each subsequent unit identical to the first one displayed on the CRT. If the system has a symbol or module library, the unit is gauranteed to be of the same uniform quality for every job.

These lower cost workstations can now make it profitable to mechanize applications that previously were considered borderline CAD applications in the mini-computer environment. Easily produced and relatively simple schematics, illustrations and viewgraphs are applications that do not require complex CAD systems and can be produced in the PC environment. It is not economically justifiable to purchase a complex CAD to meet just these applications. However, at the same time, there is nothing wrong with utilizing an existing large CAD system for these applications in non-peak hours.

A major caveat should be injected at this point. Not all low price CAD systems, even within the same price range, have equal capabilities. Within the $500-$2500 range, where you are purchasing a software package for a PC, functionality is dependent on the package design criteria. This is also true in the $30-$55,000 price range. Wide general design capabilities and object intelligence are not available until you approach $300,000 plus CAD systems or those residing on a minicomputer. An IBM PC or XT based-CAD can be used to develop drawings of approximately 1,500 to 2,000 elements, while complex drawings of up to 10,000 elements can be easily handled on VAX-based systems offered by Intergraph or Medusa operating on a Prime Computer.

A general rule of thumb that has been applied to the purchase of a CAD is the 80-20 rule. You can get 80% of the functionality of a larger CAD system for about 20% of the cost of the larger CAD system. This rule has continued to be applicable because as new complex abilities are incorporated on the mini-computers, designers are developing ways to migrate the function downward to the PC. At the same time, the resulting compressed function serves as a building block for CAD ability expansion on the larger system.

The first generation CAD applications were developed using monochromatic storage or refresh CRT's. As raster terminal technology advanced and the ability to generate color displays with reasonable resolution became available, art and animation applications migrated quickly and used up the available color spectrum of the raster terminal device. CAD applications, on the other hand, generally offered eight, twelve or sixteen colors based on the assumption that most line drawing applications could not make use of that many colors. However, mechanical CAD migrated from wireframe drawings to solids modeling and quickly the demand for a larger color pallet developed, similar to the art and animation requirements. This has resulted in display terminal ability to offer either 256 or 512 simultaneous colors in the $300,000 plus workstations and terminals such as the Tektronix 4115B.

The demand for higher image resolution has increased with the downward migration of CAD into the PC environment. By using interlaced signals and bit plane mapping, resolution has now become a function of terminal cost. The same is true of color pallet availability at the low end of the cost spectrum for CAD systems. The resolution (picture quality) and dynamic capabilities of raster systems are continuing to increase, while the costs are decreasing, thus dissipating any of the advantages formerly associated with storage and refresh displays. This has resulted in the sales of color raster displays sales growing at a rate of about three to four times faster than either storage or stroke refresh displays. The increased capabilities of the less expensive terminal have also made the migration of CAD systems to the PC attractive.

Two Case Studies

Case Study I - Implementation of Standalone CAD Workstations and PC-CADs

Case Study One is representative of an architectural firm which has increased in size rapidly over the past several years. And as mentioned earlier in the introduction, productivity costs began to play a major role in preparing technical drawings and job layouts.

Since this firm was dealing with several companies and the government, each of which had multiple sites with minor variations, the architects saw an advantage of producing standard modules on a CAD system. From the standard modules, they planned to create a full architectural layout and site plan. The government agency suggested a CAD vendor they had used for previous projects. After examining the cost of purchasing a complex architectural CAD system, the firm agonized over the $569,000 price for the computer workstation, E size plotter and software. In addition, they were faced with a three-month training period for their draftsmen to achieve full proficiency as CAD operators. This included 2 weeks of intensive training when the system was initially delivered and several follow-up workshops.

After debating whether future contracts could support the cost of an expensive complex CAD system, they chose to buy time in the decision process and to use a service bureau to meet the immediate need. It was felt that the service bureau CAD system gave them approximately a 9 times greater productivity over manual drafting methods. However, the service bureau, in addition to changing for CPU time and drafting on drawing elements and print basis, charged $70 an hour for an operator.

As a result of the 1982 and 1983 economic slow down, additional contracts did not materialize with the same companies. The architects were releaved that they had not purchased the CAD system they were initially considering. In addition, they felt the ongoing use of the service bureau CAD was extremely expensive. However, they were hooked on the productivity increases available by using a CAD system.

After attending several trade shows like the NCGA and SIGGRAPH, they talked to several vendors and were convinced that they would have to commit to spending between $40,000 to $60,000 for a CAD system to meet their ongoing needs.

After getting agreement in the partnership, they proceeded to buy a standalone graphics workstation for approximately $37,500. Basically, the workstation was a 1024 x 768 pixel, 19-inch raster monitor with 256 color display capabilities, a 32-bit processor with 4G-bytes of addressable code and data, a 4M-byte memory and a 4K-byte virtual memory to allow paging. An additional 78M-byte Winchester drive and a 1M-byte floppy disk drive were also included. The multitasking software was off-the-shelf from the vendor, therefore, a specific data symbol library had to be added by the architects. With the system purchased, there was no object intelligence capability associated with the specific symbols that were loaded into the library.

The architectural firm compared the capabilities of the standalone workstation to that of the more expensive CAD on a mini-computer. The workstation supports Fortran-77, C, Pascal and standard utilities programs and was able to perform approximately 65 to 75% of the functions of the mini-processor CAD. Due to processor speed and capabilities, they achieved only 4.4 to 1 productivity increase over the manual methods. This compared to a 9 to 1 productivity increase using the service bureau mini-based CAD, which was in the $500,000 plus price range. They felt that the lower cost compensated for the lower productivity increase.

The case study does not end there even though it makes a good case for low priced CAD workstations.

With the economic recovery in full swing, building starts increased. Therefore, the volume of the layout work increased. However, it did not increase enough to support the purchase of a second or a third standalone 32-bit processor workstation, let alone upgrading to an Intergraph or Medusa system running on a mini-computer.

This time, the firm looked at drawing packages for an Apple IIE or the MacIntosh as well as for the IBM/XT. These personal computers were chosen to handle the smaller jobs because of the good reputation and available software packages for both companies' products.

Because an apartment complex and small industrial buildings contained between three and four thousand drawing elements, it was felt that neither Apple computer could produce drawings approaching the 4.4 productivity increase achieved on the standalone CAD.

However, a new CAD package allowed the IBM/XT to approach a 3.9 productivity increase. It had approximately 80% to 85% of the functions contained on the $37,500 standalone workstation which equated to approximately 50% to 55% of the capabilities of the $569,000 mini-computer based system they initially studied.

In all fairness, eighteen months had passed since the initial evaluation of the mini-computer CAD, and the vendor had introduced a faster enhanced CAD using array processors added to the mini-computer for additional speed and expanded capabilities.

The total investment for the CAD software and the IBM/XT was $19,000. This replaced the need for hiring an additional architect and a full-time CAD operator.

Case Study II - Implementation of a Third Generation CAD System

During the late 1970's, AT&T Communications, formerly AT&T Long Lines, introduced a CAD graphics tool for the Engineering department known as the Telephone Office Planning and Engineering System (TOPES).

TOPES was an interactive computer aided design (CAD) system developed by Bell Telephone Laboratories (BTL), Western Electric Company, and Long Lines personnel to assist office planning engineering in the layout and design of telecommunications and computer equipment offices. It was also used to size and layout administrative space requirements, support systems and building elements. TOPES is not a commercially available system. At the time of initial development, because of limited display capabilities, TOPES was developed around storage tube technology to accomodate the large number of drawing elements required to support space planning. However, with the availability of high resolution refresh and raster scan terminals, TOPES achieved device independence and could be accessed from a variety of terminals. The TOPES software utilizes Plot-10 software for its basic graphics subroutine package, and for object manipulation and transformations, however, this software had been greatly enhanced by BTL. TOPES operates on multiple DEC 10 computers; therefore, a large number of users could be supported without response time deterioration.

As a result of the divestiture of the Bell System, AT&T Communications took full responsibility for supporting the Space and Power Engineers requirements. TOPES at the time of divestiture contained approximately 7,500 drawings (equipment office layouts). It is expected that the requirement for CAD generated layouts will triple to approximately 22,000 drawings of multi-function equipment floors by 1987. Some drawings contain over 7,000 elements backed up by object intelligence.

Even though TOPES had undergone several major revisions, it was felt that it would not be able to support the users' demands. In addition, a more modern, faster processor should serve as the engine for the system. Other factors contributing to the development of a new system are AT&T support of the Initial Graphics Exchange Specification (IGES) and the Graphics Kernal System (GKS), which corresponds to the Applications Layer (7) of the ISO Interface and Protocol Model.

It should be pointed out that the Space and Power Engineer requires object intelligence for each equipment frame to complete office environmental studies such as HVAC requirements, power drains, heat dissipation, floor weight and point loading studies. In addition, this object intelligence is used to complete "What-if" design studies. This could be considered a Computer-Assisted Engineering (CAE) application.

TOPES, in the present form, has a productivity factor of 5.6 and provides a $6.5 million savings.

The development of the Multi-user Engineering Graphics System (MEGS) to replace TOPES will increase their productivity factor to approximately 8.0 and the savings will increase by $3 million to a $9.5 million annual savings.

The 55 potential users have approximately 23 terminals located in 7 different locations. The associated object intelligence database is too large to replicate on multiple standalone workstations and must be available to all locations simultaneously.

The individual office drawings are used by only one engineering location, and the AT&T Headquarters group must be able to access all the drawings to conduct various studies. However, these drawings can reside on disks associated with regional processors and centralization is not required.

In this case, neither standalone workstations or PC-CADs would meet the user requirements. Therefore, a star network of mini-computers is being recommended to meet the business needs of the 1987 environment.

Conclusion

The availability of low cost standalone CAD workstations and PC CADs are making a valuable contribution to productivity in various functional disciplines. These low-cost CAD systems are generally developed on multitasking microprocessors or PCs using standard operating systems. There is an increasing number of general purpose drafting software and application specific packages that can now be used for design applications that in the past, would not economically support mechanization on a mini-computer based CAD system.

The increased intelligence and power available by using chips like the AT&T Technologies WE32000 or other 32-bit chips from Motorola, Intel or Hewlett-Packard in the computer design allow complex functions to be run on smaller processors. When you examine the current marketplace, 40 new low cost CAD products were introduced over the past two years. And the trend will continue.

Based on some observations made by Carl Machover, by 1986 the low cost CAD systems will capture approximately 20 percent of the total CAD market. This equates to approximately $800 million out of a total $400 billion market.

This trend pojection is supported by a change in product strategy by corporations such as Computervision. Computervision has introduced a low cost 2D drafting workstation that is expandable with entry workstations priced from $35,000 to $52,000. Other CAD vendors are following suit.

As the two case studies indicate, there are applications that will continue to require dedicated mini-computers. However, the trend shows a definite migration to a low cost workstation environment or independently developed CAD software for the PC.

Biography

Mr. Bannon is the Manager - Systems Engineering Computer Graphics Consultant for AT&T Communications. He is responsible for planning and implementation of plans that direct the integration of Computer Graphics into the Engineering EDP environment. These plans incorporate new technology and concepts into the corporate decision analysis support process. He is responsible for all phases of computer graphics - CAE, CAD, CAM and Business Graphics. This includes earth science cartography systems.

Mr. Bannon studied for a BSEE at Pennsylvania State University and is a member of the NCGA, IEEE, IEEE Computer Society, and SAME. He is a member of the Advisory and Program Committees for CADCON and ATI. He has participated both nationally and internationally in conferences such as CAMP, MICAD and CG Tokyo and he has authored over 20 computer graphics papers.

4

Advanced Systems Integration Management Via CAD Graphics

Robert T. Bannon
Manager-Systems Engineering
Computer Graphics Planning
AT&T-Communications
Bedminster, New Jersey 07921, U.S.A.

Abstract

The purpose of this paper is to explain the creation of an Electronic Data Processing Architecture using a Computer-Aided Design (CAD) System. The EDP Architecture includes the interrelationships and data dependencies between EDP systems required for effective systems integration management. For purposes of illustration, this paper will use the functional relationship required to support telecommunications engineering.

By using computer graphics to define EDP architecture, the following advantages become evident:

- Graphic mapping of interfaces inform EDP systems designers of user requirements. An example of this is the access of equipment requirements for determining floor space requirements.
- Inform both users and upper management of proposed EDP design activity.
- Color display and output capabilities provide the capability to highlight major impact items.
- Graphics structure allows the EDP Architecture to clearly convey the maximum information required for a specific application function.
- Dynamic graphics teleconferencing allows for immediate viewing from remote locations.

A suggested methodology of preparing an integrated EDP Architecture using Computer Graphics is described and can be applied to other technologies and industries.

Introduction

In a Booz-Allen study conducted in 1983, it was estimated that almost half the United States workforce was engaged in information processing. Since that study was conducted, information processing requirements and computing graphics output of this information has grown rapidly. There is a increasing number of companies and research institutions becoming more deeply involved in data manipulation, processing and information management. Because of this rapid growth, it is not uncommon to find that similar or redundant functions are being performed in various departments. In extreme cases, EDP systems within the same department may even store and manipulate the same data. This is caused by poor EDP system design, changing user requirements, parochial interests or proprietorship of information. Organizations do not want to disseminate their knowledge if this causes additional work or loss of total control of their system.

System integration architecture is a complex planning procedure that does not lend itself to analytical treatment of its effectiveness. Therefore, to plan system integration and an efficient data flow between systems, with a minimum of human interfaces and functional duplications, the establishment of an EDP Planning Organization is required. This group can reside within either the user or design organization. Its function is to create an integrated EDP System Architecture which depicts both the functional work flow and the interrelationship between systems. The basic thrust of the EDP Architecture is to provide an automated coordination and mechanization of interfaces for better data consistency and information management.

This paper discusses the use of a Computer-Aided Design (CAD) System to create the EDP system integration layout and to manipulate the EDP Architecture. By using the CAD system for these purposes, we are able to map the changes in information developed in one system and passed downstream to another. Also, we are able to convey, with a high degree of clarity, major impacts on the EDP system architecture. This in turn results in better system design for a new system or set of programs that will be integrated into the overall functional work flow.

CAD System Selection Criteria

The selection of a computer graphics system to support the design of an EDP system architecture is similar in nature to selecting a graphics system to support either circuit board or building design.

Human interfaces with the graphics system is via a workstation. Today's workstations are ergonomically designed to relieve eye strain and body fatigue from working for prolonged periods. The operator creates, modifies, and refines the architectural layout interactively, viewing the emerging work on a graphic display. With a single command, the architectural designer can manipulate and change the proposed data flow highlight impact areas in the output format or change the relational view.

A typical workstation will consist of a terminal (CRT), a hardcopy device and a variety of peripheral hardware that enhances a system's flexibility and responsiveness the EDP architectural designers need.

To construct the EDP architecture and map functional relationships, you may choose to use either an existing in-house computer graphics system or opt for a new system. When using existing hardware and software, the initial economic outlay is less than that of a new system, however, there are some trade-offs that must be considered:

- What is the volume of work already residing on existing graphics systems?
- Is the processing environment a Mainframe or Mini-computer or workstation micro-based?
- Is access time readily available, or is it on a special priority or allocation basis?
- Is the workstation convenient for the architectural designer, such as in a common terminal area? Or will access be dependent on purchasing a new separate workstation?
- Does the existing workstation design meet the user requirements of the architectural designer without modification? And if modifications are required, how extensive are they?
- Is the software in a programming language, such as Fortran-77 or "C", supported by the potential designer organization? If software changes are required, does it require the vendor to make changes?
- Does the system support the Initial Graphics Exchange Standard (IGES) and the Graphics Kernal System (GKS) standards?
- Was the graphics system designed to be menu driven only, or does it have the flexibility to allow the creation of special symbols, line manipulation and alpha-numeric text input and editing capabilities?
- Is existing documentation sufficient to allow the existing CAD system to be manipulated for EDP Architectural integration?
- What are the operational costs of the existing system?
- How extensive is the EDP Architecture you intend to create? Does it support only one department, or does it encompass the entire EDP resources of your entire company?
- Will your EDP Architecture include only the function and system level? Or will it branch into schemas and subschemata for functions and systems?

After investigating the capabilities and limitations of your existing CAD system you may decide that it will meet your requirements or a majority of them effectively. If there are legitimate concerns about its efficiency and usefulness you may want to investigate a new system.

It should be remembered that most CAD graphics systems are designed to allow interactive manipulation of symbol location and associated text. Business graphics systems and statistical graphic packages do not have the flexibility required for this aspect of EDP architectural design, therefore these systems are unsuited for an EDP architectural design application.

In investigating the purchase of a new graphics package to support the design of an EDP architecture, some additional considerations should be evaluated:

- Total package price?
- Cost/performance comparison?
- Processor features - Micro-, Mini-, or Mainframe?
- System Capacity?
- Interface Compatibility?
- Quantity of workstations system can support?
- Expandability?
- System performance specification?

CAD Selection

Based on an evaluation of our Engineering Department requirements, the time constraints associated with introducing a new graphics system and the versatility of an existing system, we choose to initially implement the EDP Network Systems Architecture on the Telephone Office Planning and Engineering Sytem (TOPES). Since the original decision to use TOPES, we evolved into replacing TOPES with a third generation CAD system, Multi-user Engineering Graphics System (MEGS). As a spinoff to MEGS, we are developing a UNIX, System 5.0-based CAD graphics application to support the EDP architectural design. This application is known as the Mini Unix Graphics System (MUGS).

TOPES was an interactive computer aided design (CAD) system that was developed by Bell Laboratories (BL), Western Electric Company, and AT&T Long Lines personnel to assist office planning engineering in the layout and design of telecommunications and computer equipment offices. It was also used to size and layout administrative space requirements, support systems and building elements. TOPES was not a commercially available system. At the time of initial development, because of limited display capabilities, TOPES was developed around storage tube technology to accommodate the large number of drawing elements required to support space planning. However, high resolution refresh and raster scan terminals allowed TOPES to achieved device independence and it can be access from a variety of terminals. The TOPES software utilized PLOT-10 software for its basic graphics subroutine package, and for object manipulation and transformations, however this software was greatly enhanced by Bell Laboratories. TOPES operated on a DEC 10 computer, therefore approximately 16 concurrent users were supported without response time deterioration.

The development of MUGS will allow the EDP Architecture to migrate from a 36 bit processor to a 32-bit environment and to utilize new workstation features that have become available during the second half of 1984. In addition, the PLOT-10 graphics core is being replaced with the new version of Precision Visuals, DI3000 which will be Graphics Kernal System (GKS) compatible. MUGS runs on either a DEC VAX 11/780 or the AT&T Technologies 3B20 processor. MUGS use INGRES as its database management system (DBMS).

As with my company, time constraints will weigh heavily upon your evaluation and will influence your decision to use an existing in-house system, permanently or for an interim period, or opt to implement a new system.

EDP Systems Architecture

EDP System Architectural Design is a discipline that evolved out of a combination of engineering and EDP program documentation criteria. It is based upon the CAD technology that evolved from schematic drawing layout for printed circuit boards and building design. At the same time, EDP architecture utilizes techniques associated with data flow and system documentation. The motivation for developing an EDP System Architecture is the creation and documentation of a logical model of system interrelationships for a functional or engineering technology. By using CAD graphical techniques it enables engineers, users, analysts and designers to develop a concise picture of the various system dependencies and data access requirements existing between functional groupings of systems.

Ironically, system designers and programmers have relied on manual drafting techniques for system documentation, even while they were designing computer graphics systems. The use of a CAD system has eliminated this manual bottleneck in developing and publishing an EDP Systems Architecture. These non-creative and error-prone procedures can be costly and contribute to the untimeliness and inaccuracy of an EDP Architecture which must be a dynamic transition document.

Implementation of a CAD system in the EDP planning environment has eliminated much of the frustration incurred while making changes in a complex EDP Systems Architecture which contains a maze of engineering support systems that integrate functionally on information processing levels. Manual modifications that reflect system interconnection can be tedious, but with computer graphics the process will be expedited and the EDP Architecture will become a dynamic tool that can reflect transitional impacts from subschema level through functional level in a timely fashion. This allows the EDP Planning Organization to answer "what if" questions concerning downstream impact if the generation of certain data is eliminated or the format is modified.

With the old method of manually drawing system interrelationships, the input was usually a hand-drawn sketch prepared by the EDP planner, which was then redrawn into an understandable format that met corporate standards. This included the drawing of each system symbol separately, interconnection of these symbols based on the hand-drawn sketch, and finally annotating flow lines and placing an EDP system name within each symbol. Allowing for correction of errors and changes in subschema flows as projects become more defined, this process went through several reiterations before final distribuiton. This was a very time-consuming task, requiring constant interaction between system designers, EDP planning staff and the drafting department.

The use of the CAD system should not require the user to be familiar with a programming language, however our EDP planners have both an engineering and programming background. Therefore, the introduction of CAD system into this process was an easy transition. The need for the development of a rough sketch and the drafting department translation into a corporate format that underwent multiple revisions were eliminated. The EDP Planner now creates an integration architecture directly in the CAD system. Our CAD system has become the electronic scratch pad for the EDP planner, and consistently produces high-quality drawings. The transition to MUGS will allow an expansion of the user community because the command language can be either at the conversational level or at a more advanced level for frequent users.

The creation of an EDP System Architecture is usually created by repeating the following steps for functional engineering clusters:

- Create a drawing grid or call up a standard grid from a symbol library, including border and legend generation.
- Placement of individual system symbols in functional clusters on the grid. Symbols are drawn once and stored in a library from which they can be recalled from a menu or by a simple command.
- Interconnection of system symbols. This is accomplished with line segments and arrows used to depict direction of interface flow or by use of connector symbols which are drawn adjacent to each system symbol. As a drawing convention, the connectors are used when the interconnection crosses three interface lines or 5 horizontal units. Each connector symbol uses and alpha-numeric code, i.e., A or B2, and is recorded on an interface connector table for reference ease.
- Reposition systems for functional cluster interface or interface with other systems.
- Annotation or printing text on the EDP Systems Architecture.

CAD Library and Symbol Retrieval

The establishment of an EDP-oriented symbol library is one of the major factors contributing to the cost effectiveness and dynamic flexibility of using a CAD system to develop EDP architectural interrelationships. The storage of frequently used symbols greatly simplifies the preparation of the EDP Architecture drawing. These symbols may be simple, such as an arrowhead or extremely complex, such as the EDP architectural symbol for a mechanized non-engineering interface process, which includes border shading. Once these symbols are created and stored in the library or database, the symbols can be recalled and incorporated into the drawing at the proper coordinate points. The method of retrieval is dependent upon the terminal hardware employed.

The use of mouse in conjunction with a menu is synonymous with current CAD technology. This peripheral input device allows a symbol to be inserted into a drawing in seconds with just the touch of a button. Prior to 1982, systems based on raster terminal output were limited to fairly simple displays because of limited screen resolution. However, availability of high resolution color displays (up to 1240 x 1240) at reasonable prices have now allowed complex drawings to be presented.

With TOPES, the initial drawing symbol was added on the drawing using a four-letter command after the terminal crosshairs are positioned at the desired X- and Y- coordinated. MUGS will use a menu and mouse to place symbols on the drawing. Symbols can be moved and relocated via the mouse or the symbol can then be copied at other drawing locations by repositioning the crosshairs to a new set of X- and Y- coordinates via the thumbwheel cursor controls or via a joystick and then issuing a single letter command (i.e., "c") to copy.

CAD Input

The EDP Planner can design the architectural layout "on-line" or within a local, down-loaded micro-processor which stores the drawing coordinates and associated symbols on either a floppy or Winchester disk for block transmission to the host processor. The drawing input capabilities are inherent to the CAD system selected. Both modes of operation allow the system to respond immediately when compared to manual methods. A major advantage to the introduction of CAD systems into this area of EDP engineering is that as the architecture and associated data is being stored in the processor, visual feedback is drawn on the CRT which allows immediate on-line editing and rapid error correction.

When creating this type of drawing, many system symbols are the same, however, system titles are different which require the individual key entry of the associated text. Standard annotations, such as flow notes and test strings, can be stored in the library for recall. The test string can then be placed at the appropriate X- and Y- coordinates via the cursor/command sequences already described.

EDP Architecture Functional Geometry and Integration

CAD systems use the computer's processing capabilities to eliminate many of the problems associated with the placement and interconnection of the various symbols utilized in an EDP Architecture drawing. However, for these features to be utilized the EDP planner must establish a placement grid and a functional geometry.

Once the grid is established, the EDP planner functionally divides the grid for system placement. By functional division, it is meant that an area that serves and engineering discipline is reserved on the grid. For telecommunications engineering these functional disciplines might include:

- Network Planning
- Switching Planning
- Equipment Planning
- Internode Planning
- Facility Planning
- etc.

To establish an EDP System Architecture for other than telecommunications technology, substitute the appropriate engineering function or industrial responsibility for the functional division. An advantage of the functional division is that it allows the establishment of boundary coordinates for "zooming-in" on a portion of the overall Architecture. This allows small areas of the drawing to be copied in an enlarged state for the purpose of clarity or if higher management's interests are focused on impact areas within a specific functional discipline.

Once the functional strategy and grid geometry are established, the various system symbols, i.e., network engineering symbols or non-network interfacing system symbols, are placed on the drawing in the manners previously described. The system symbol placement utilizes a strategy to minimize total length of the interface connections and limit line bridging once the systems are interconnected. This "mapping" technique is one of the most difficult for the EDP planner to apply. The planner must take into account the various drawing standards and physical constraints imposed by the functional grids and engineering clusters. Using the CAD system, the EDP planner must arrive at a functional placement that optimizes subsequent interface routes on the drawing without creating an undecipherable congestion. This is a matter of experience that may require multiple iterations, however, based on PC board and cartography applications, the use of a CAD system has a potential of a 5 to 1 time savings over manual drafting.

Some PC board graphics systems are equipped with an automatic router software which can be adapted for preparing the EDP Architecture. This is accomplished by assigning a location and zone to a symbol, similar to gate assignment. For example, B7R3 to D4T1 would mean to connect the third right zone of the symbol in "B" vertical, seventh horizontal to the first zone on top of the symbol in "D" vertical fourth horizontal. However, if your system is not equipped with an automatic router, and the input method is by terminal input, the interconnections are drawn by cursor control/command methodology.

If the CAD system selected for creating the EDP Systems Architecture utilizes either refresh or raster technology, there are several features that can be utilized to expedite drawing the interrelationships without abnormal congestion. Once the systems (symbol and system name) are placed on the drawing, a PC board design feature referred to as "RATS

NEST" can be utilized. This feature allows the computer to determine interconnections paths and provides the EDP planner with a graphic representation of the proposed mapping. From this graphic display the planner can determine whether a system must be moved to provide a cleaner interconnection route. At this point in the Architecture layout, the EDP planner can activate another PC board graphics design feature, "Rubber-banding." The planner places the graphics system into a dynamic mode which allows the EDP system and symbol to move or track across the screen and positions the symbol at potential locations. While tracking, the system interfaces stretch or bend while conforming to drawing standards, i.e., lines do not cross symbols but route around them. This gives the EDP planner the opportunity to select the optimum location. An adjunct standard that could be added to the "Rubber-banding" feature is the use of drawing connectors. If the interconnection line is forced to bridge across 3 or more system interconnections, or if the interconnection must pass thru 5 or more horizontal grid areas, then the CAD system will default to the automatic selection of the next available connector, i.e., B2. The B2 connector would subsequently be placed adjacent to both the upstream and downstream systems and arrows would be drawn to show flow direction.

Editing and Flow Checking

Throughout the layout process, it is necessary to perform various editing functions. The editing features are powerful commands that are incorporated in well designed CAD systems, whether they run on storage tube, refreshed or raster terminals. Interactive editing functions include commands such as layer or overlay discrimination and transfer. They may also include a color highlighting or a glow command which allows the EDP planner to highlight major impact areas and systems. Another editing feature is the ability to change line type and weight, i.e., solid, dotted or dashed, heavy, etc. This feature conveys a high degree of clarity and information about a specific function. An example of this would utilize one line type and weight to show an interface between two IBM mainframe based systems, while another line type and weight could be used for a proposed interface between a system run on a AT&T 3B20 processor and one run on a DEC mini-computer.

Additional editing commands exist that control functions such as moves, deletions, text edits and symbol and job renames.

If the EDP Systems Architecture is driven by an associated database, then a base exists to check the interconnections between systems. This can be approached either by manual inspection, or if your system is so equipped, by using the data extract process along with the default or "not-equal-to" linkage to print discrepancies.

At the same time, when a new system or interface between systems is added the database design can flag additional changes and deletions and generate a CAD input report. For our non-graphic telecommunications engineering systems database we use RAMIS as our data manager and report writer. This was employed in conjunction with TOPES, however as stated earlier, we will be using INGRES as our DBMS with future developments.

Color Enhancement

With the availability of high resolution raster workstations (1240 x 1240), color becomes an important feature for highlighting many different types of data. When creating an EDP Architecture, color can be used to show the telecommunications functional disciplines, i.e., Equipment Planning, system developmental status, Systems Management Organization (SMO) or even annual operational costs.

Advances in color output devices, coupled with the device independence attainable through GKS conversion, allows our EDP architecture to be outputted to a wide variety of devices. Initially, color output was limited to either pen plotters or expensive film recorders.

Cost effective color output is now available from the Seiko Color Hardcopy unit, which uses thermal ink transfer technology, and low cost ink jet plotters from Tektronix, ACT and Diablo. In addition, inexpensive screen cone systems are avialable for color slide production from either Kodak or Polaroid. These are usable for peer presentation.

Good resolution, presentation quality slides can be processed on the Samuri and Polaroid Palette systems. These units are far less expensive than slide generation systems of 2 or 3 years ago.

Graphics Teleconferencing

If the CAD system is run on a mini-computer as its only processor, between 4 to 16 remote terminals is the maximum configuration. This limitation is a function of engine size, access and processing capabilities. One advantage of a mainframe environment is that there can be a significantly larger number of available ports for individual users access. However, depending on the number on concurrent users, response time may deteriorate to an unacceptable level. Drawings can be downloaded to microprocessors located at the workstations for local graphic processing and transferred to the mainframe for drawing archival.

One of the most significant features of our CAD system is that of graphics teleconferencing from remote locations. Depending on port and conferencing bridge capabilities of the operating system and the CAD systems, it is not uncommon to have 4 to 16 simultaneous remote conferees watching the same graphics output. By design, manipulation of the drawing symbols can be relinquished to remote locations, however the initiating drawing control ID maintains the ability to issue permanent overwrite or storage commands.

An advantage of graphics teleconferencing is that an EDP planner can be at a remote location and still make a presentation to upper management who may be at a remote management information and monitoring center.

An enhancement that we are investigating includes linking color code capabilities and subschema logic. The graphics teleconferencing when coupled with color capabilities and editing features allow "what-if" questions to be answered in an interactive mode. For example, if the long range (beyond 5 year) circuit input requirements from a forecast conditioning system were eliminated, what is the resulting impact on Space and Power Planning?

To answer this, the current path is depicted in one color. Analysis of functional schema tasks performed in various Equipment Planning systems show that the circuits must be converted into orderable equipment quantities which are then converted to full bay equipment compliments. The full bay compliments are the basis for Space and Power Planning.

When the deleted input is shown, the subsequent functions glow or flash repetitively to indicate an engineering function jeopardy. The functional schema and subschema level must be searched to see if a modification or change can produce the required data. If it can, the "RATS NEST" feature can be activated to map the new interconnection path. When this is accomplished the line color, type and glow frequency can be changed to indicate the answer to the "what-if" question. The EDP planner who has the drawing control ID will determine if the new input should become a permanent part of the EDP Architecture.

Conclusion

As in many technical fields, the design of Telecommunications EDP Systems Architecture requires multi-discipline expertise. This position requires a strong technical background in engineering function as well as an understanding of EDP System design and development. The need for people with these special skills is consistently outpacing the personnel available. Examination of business opportunities are frequently postponed until an EDP Planner is available to address the EDP solution and incorporate it into the Architecture. This is even more apparent if the entire planning process is manual and computer graphics are not applied.

Therefore, it become essential that the EDP Planner's knowledge be recognized as a resource that can be enhanced with a CAD system and result in a productivity increase.

The advantages of introducing a CAD system into the EDP Planning environment are increasingly obvious. The CAD system rapidly produces an overall view of EDP systems interrelationships and functional dependencies.

The CAD system enables a human-computer synergism to exist. The system supports the creative abilities of the EDP Planner and the interaction of the two produces an Architecture that neither is capable of producing alone in a timely fashion. Clearly, a CAD system can produce a significant time savings over manual methods.

Additionally, an up-to-date Architecture can provide a method for tracking EDP development projects and map their overall impacts. This keeps both the user and upper management informed of EDP activity.

Color graphics and editing features inherent to the CAD design allow the Planner to highlight major impact items and perform architectural "what-if" studies if a subschema level of detail is available.

If the CAD resides on a processor that allows remote access another benefit of using a CAD system to develop an EDP Architecture is the dynamic graphics teleconferencing capability which allows immediate viewing from multiple locations.

CAD systems and applications are constantly being improved and enhanced and we have not yet discovered their technical limitations. The only suggested limitation is the human imagination. EDP Planners should be at the forefront of CAD system usage and take advantage of the mechanized capabilities they have created for other functions and technologies.

Biography

Mr. Bannon is the Manager - Systems Engineering Computer Graphics Consultant for AT&T Communications. He is responsible for planning and implementation of plans that direct the integration of Computer Graphics into the Engineering EDP environment. These plans incorporate new technology and concepts into the corporate decision analysis support process. He is responsible for all phases of computer graphics - CAE, CAD, CAM and Business Graphics. This includes earth science cartography systems.

Mr. Bannon studied for a BSEE at Pennsylvania State University and is a member of the NCGA, IEEE, IEEE Computer Society, and SAME. He is a member of the Advisory and Program Committees for CADCON and ATI. He has participated both nationally and internationally in conferences such as CAMP, MICAD and CG Tokyo and he has authored over 20 computer graphics papers.

5

Developing A Cartographic Geo-Code System (CAGES)

Robert T. Bannon
Manager-Systems Engineering
Computer Graphics Consultant
AT&T Communications
Bedminster, New Jersey 07921, U.S.A.

Introduction

As a result of the divestiture of the Bell System, AT&T Communications is faced with creating and maintaining various geographic maps that would be used by organizations ranging from Planning and Design Engineering through Business Research and Marketing. Many of these applications were previously supported by the Bell Operating Companies or contracted to outside vendors.

Some of the traditional applications requiring mapped information are distribution service mapping for utility access in metropolitan areas (right-of-way), terrain mapping for determining radio route interference and tax block data. As the new AT&T Communications evolved, it became apparent that mapping support would also be impacted by the Modification of Final Judgement (MFJ) and the Plan of Reorganziation (POR). For example, maps are required to provide Local Access Transport Area (LATA) boundary information. These LATA boundaries quite often do not conform to municipal and county orders, therefore they represent another information layer.

To allow detailed office locations to be plotted for applications, such as facility planning, and the determination of specific "interchange point" connections the computer generated maps require a high degree of accuracy. To achieve this, location points must be plotted on the map using x,y coordinates derived from specific latitude and longitudes or from Vertical and Horizontal (V and H) coordinates, which in turn are derived from latitude and longitude. The same geo-coding capabilities are required for mapping customer locations.

Various organizations which were formerly served by the Bell Operating Companies started contracting with numerous vendors to meet their specific needs. This meant that there was a large amount of overlapping cartographic data that was being paid for by these various organizations. Since our corporate requirements are not the only demand on the cartographic vendors and service bureaus, delays are often encountered. This condition exists whether the vendor uses computer generated maps, manual drafting methods or acetate overlays on large scale public domain maps.

Based on the increasing demand from the various AT&T Communications organizations, System Engineering has undertaken a feasibility study to determine the most economical means of providing computer graphics support for the various user communities. The central part of this feasibility study will be based on a proposed interactive geographic and demographic information mapping and management system, the Cartographic Geo-Code System (CAGES).

Post Divestiture Environment

As stated in the introduction, prior to divestiture many of the mapping applications for AT&T were handled by the Bell Operating Companies or contracted to a vendor that the Operating Companies were using for plotting shared planning data. However, as a result of the Bell System break up, AT&T Communications and the Bell Operating Companies were limited in the amount of data that could be shared. This resulted in the various departments entering into direct contracts with vendors and service bureaus to provide cartographic output.

Since there were so many separate organizations requesting maps, duplicate efforts were being performed by different vendors and service bureaus. In fact, because of time frame requirements and parochial control issues, identical requests from several groups within the same AT&T department were forwarded to a single cartography house.

In another case, information that AT&T had supplied to a Bell Operating Company could no longer be mapped for AT&T-C by a vendor because this would have been considered sharing of company proprietary data and a violation of the MFJ.

To really appreciate the diversity of the mapping requests being generated, a few applications are described below under functional headings.

Radio Route Planning

The radio engineer uses Landsat/Geodetic information to design a radio relay signal path between two towers. To accomplish this he must know of any obstruction that would interfere with signal propagation. To properly plan the route, the engineer must calculate interference along the path based on shadow mapping and elevation adjustments because of earth curvature and possible structural impairment. The radio engineer currently uses a scattergram interference output generated by DISPLAA and overlays this on geodetic maps. Incorporation of support algorithms would result in the ability to map the route directly on the screen, and the radio route could be kept in data file with other facility maps for use in the future.

Cable Route Maintenance and Icon Tracking

Our Operations Department currently has various locations throughout the country that deal with construction contractors who will be digging in areas adjacent to our underground plant. In the current environment, a contractor calls a local office, whose telephone number is posted on a cable marker to report where he is digging. The Operations Department representative determines from the manual record drawing if it is okay to proceed with an excavation or whether an outside plant representative should be dispatched to the site. Today, two or more centers exist for each state.

Efforts are currently underway to develop a detailed facility mapping system to support this function, however this application should be incorporated into CAGES. As a result of this cartographic development, the number of reporting sites could be reduced from fifty-seven to seven or ten sites. Along with the site reduction, there would also be a corresponding personnel reduction. To achieve the user objective, the system would require an accurate detailed cartographic system that could pinpoint the exact construction location within 10 seconds. In addition, this same mapping system could be used to track construction gangs or repair trucks via the use of icons.

Customer Location Mapping

In the current environment, when an organization needs to see graphically the location of customers, they have contracted with a service bureau to plot customer record files on a map of a specific area, e.g., Anaheim, California. Their current "hit" ratio in plotting the customer location is approximately 70% accuracy for addresses fed into the service bureau computer. The remaining 30% which was originally unplottable, is resolved by manually entering additional address data into the computer or extending streets, or manually entering in new areas.

By developing or purchasing a cartography system for AT&T-C, we do not expect to increase the location matching accuracy initially. The seventy percent is considered to be an acceptable matching ratio; this is due to address inaccuracies, (street, city and zip-code), regional growth and plain old human error. However, as we build the customer data base and populate the fields for the associated latitude and longitude or Vertical and Horizontal Coordinates (V and H Coordinates), we expect approximately a ten percent increase in matches. These additional attributes can be used for various engineering criteria, therefore they are not being added just to increase the initial load ratio.

Local Area Transport Area (LATA) and Point-of-Presence (POP) Mapping

As a result of divestiture, the establishment of service boundaries or zones were established to define areas served by the various telephone entities. Both the divested Operating Companies and AT&T-C have contracted with various service bureaus to create these maps, however they must be kept up-to-date. New offices must be plotted for both the AT&T-C Point-of-Presence (POP) and the Bell Operating Company and Independent Local Serving Offices (LSO). This should also include feature group (type of service provided) information.

It does not make sense to have another company maintain our installed location information and our company proprietary planning data. This is especially true if their internal company and computer security standards do not conform to ours.

Indigenous Customer Studies

Our Business Research Department conducts studies for various organizations, such as Marketing, that identify customer grouping in a particular area.

For example, if you looked at the pharmaceutical industry around the Philadelphia, Pennsylvania area you would get one type of scattergram to overlay on an area map. However, if you expand the scope to include other chemical producing customers, you get a different perspective. And if you expand the scope to include petro-chemical companies, such as DuPont and Atlantic-Richfield (ARCO), you get a third distribution map.

A cartographic-demographic system linked to a customer data base would lend itself ideally to this type of application. The inclusion of color output would portray the information required for planning in a readily identifiable format.

Proposed System Functionality

It is assumed that any system introduced to support the AT&T-C cartography application would be "state-of-the-art." Although this phrase is overused, in this case it can be interpreted as meaning that the system will utilize an open architecture, include the latest features, support graphics standards and support a wide variety of work stations. This ranges from the AT&T-C Interactive Planning Work Station (IPWS), the Tektronix 4105, 4107, 4109, 4114 and 4115B to the IBM PC and XT environment. This can be achieved by a micro-coded "termcap" that replaces device drivers at the front end of CAGES. The termcap is easily updated to include new terminals and devices as they become available. Short of this, an emulator package for a specific terminal could be used. Along with utilizing raster terminal technology, the system would support a full-color pallet, 256-color minimum, and utilize a 32-bit map data base interactively.

Since the AT&T-C operating area is nationwide, the size and precision of the graphics data base must allow the construction of a continuous base map from assorted original map sources. When constructing an integrated map from individual map segments, generation of a standard scale and storage format is essential. Both marker (latitude and longitude reference points) and edge matching procedures will electronically align the map segments while minimizing distortion.

The sources of geographic data used to create the initial continuous map digitization will come from a variety of places:

- Landsat Photogrammy (Satellite Photographs)
- Geodetic/Topographical Maps (Topos)
- Electronic Theodolite (Electronic Survey Coordinates)
- GBF/DIME Files (U.S. Census Bureau SMSA Mapping Data)
- STAMAP (State Planear Maps)
- Manually Digitized Maps (Mouse Encoded Geographic Data)

By using the transformation capability inherent to a specific graphics core, such as DI3000 or Tektronix IGL, and a least squares-elastic body algorithm, the system will have the capability to merge a variety of map segments, differing in size and scale, into a single, continuous map of the United States. A second transformation process would match various map grid systems to a latitude and longitude base which in turn would be stored as x,y coordinates for future manipulation.

Zoom and pan features along with the application of Boolean polygon generation, will allow the zooming-in on an irregular polygon area such as a state, city, zip code or a several block area. The polygon attributes can be defined in the same manner as those in the SMSA files and include tax block lot and address attributes.

The system must allow for the integration of new Theodolite or Photogrammy data into the geographic data base as well as tape file update and manual input. This would require a stereodigitization capability that would probably be contracted to regional survey corporations. This brings up the point of input standardization, which will be discussed under the next heading. However, it must be pointed out that the contracted format must match the Initial Graphics Exchange Standard (IGES) for transporting CAD data, the Graphics Kernel System (GKS) which defines 2-dimensional objects and the Virtual Device Metafile (VDM) which represents a standard file format for 2-D graphics data.

These same standards would apply to data captured by various scanning techniques. These techniques range from line followers through digital cameras coupled to optic fiber scan heads. Basically, this technology scans existing maps and plats (outside plant drawings for telephone plant and other utilities) to derive a rasterized image which is then converted to a vector format. By assigning the latitude and longitude of interlock markers on the drawings these can also be integrated into the geographic data base as x,y coordinates.

Map discrepancies, resulting from either incomplete digitized information from the file merger of the GBF/DIME file and automated conversions of other maps or from human input entry, must be corrected by human interface using a workstation. The engineer or cartography system operator compares proof plots to the original documents or files, and verifies the accuracy of graphics-coded representations. In addition, the engineer runs polygon closure routines, based on Boolean applications, which can identify any polygonal inadequacies. Interactive map editing capabilities must be provided to allow for boundary modifications, corrections and additions.

Overlaying capabilities are required that allows the system user to view an existing segment of the cartographic file or area map and superimpose geographic or demographic models for quick comparisons and "what if" study capabilities. This feature is similar to those incorporated into the Multi-user Engineering Graphics System (MEGS) which is a third generation space planning and architectural CAD system also being developed by AT&T-C. To increase the system flexibility to meet the modeling and "what-if" analysis overlay demands, linkages must be built to non-graphics data bases to allow information that is crucial for analysis to be extracted and included on the cartographic model. Once again, this points to the need for creating CAGES in an IGES format.

Maps and related icons, such as cable vaults, repeater symbols or customer-industry symbols, must have both defineable and scale dependent symbology default capabilities to allow for specific relevant details to be drawn for analysis.

The development of a thematic analysis package would allow interactive analysis of spatial relationships between complex data sets using a variety of quantitative and statistical management techniques. Non-graphics analysis (mean, standard deviation, population variance, trend, linear regression, etc.) as well as polygonial analysis (intersection, union and subtraction logic) should be developed. The resultant data set represents the multi-parent relationship between data file entities and accurately portrays complex logic relationships without redundant storage. This in turn can be used for further graphics manipulation analysis such as zone generation, point and icon overlays and common boundary removal. The calculated results from thematic analysis should be retrievable via cartographic output or via an alphanumeric report extract. Since the type of analysis will vary greatly between the various user communities, custom reports as well as standard reports will be required, therefore, a report writer linkage, such as the RAMIS Report Writer should be included in the system development.

Since outside industry is migrating to voice recognition as an input capability, considerations should be given to this technology as a system input method as well as mouse/menu input. Voice recognition systems have developed from 10-words "untrained" vocabularies to well over 100-words "untrained" vocabularies during 1984. "Untrained" means that it will recognize specific commands from a variety of users without the system undergoing a "user-voice-specific-familiarization" training session.

Computer Graphics Standards

Standardization clearly effects the development of graphics software packages much the same as other types of computer applications. It reduces development time and simplifies programming. Standards avoid the need for constantly reinventing the wheel by allowing a graphics drawing package or "core" to incorporate common computer graphics functions in a predefined format. Adoption of standards translates into reduced development costs and provides the ability to incorporate various vendor software packages into a graphics application.

Various graphics committees have issued guidelines as an attempt to regulate or define computer graphics formats. The committees recognized by the National Computer Graphics Association (NCGA) and the World Computer Graphics Association (WCGA) for standards development are:

- American National Standards Institute (ANSI)
- Institute of Electrical and Electronics Engineers (IEEE) Computer Society
- International Standards Organization (ISO)
- Y-14 Committee
- Independent Software Information Standard (ISIS) Group
- National Committee for Digital Cartographic Data Standards (NCDCDS) - part of the American Congress on Surveying and Mapping (ACSM)

End users are the driving force in making the various hardware and software vendors conform to computer graphics standards. They have a definite need to have various systems and data bases interface.

Many companies try to adhere to standards, others adhere only when it suits their company's purpose. One attitude of turnkey vendors is that if a user has data files written using the graphics Kernel System (GKS) file structure, they'll provide a utility to translate that information. However, they are not going to make files compatible to run on a competitor's machine. Software vendors take another stance and depend on their ability to create device independent programs or to develop drivers for industry leading equipment. General acceptance of standards promote the portability of their software to a wide variety of processors.

The following figure shows the relationship between the various standards that are required to develop a versatile cartography like we are proposing.

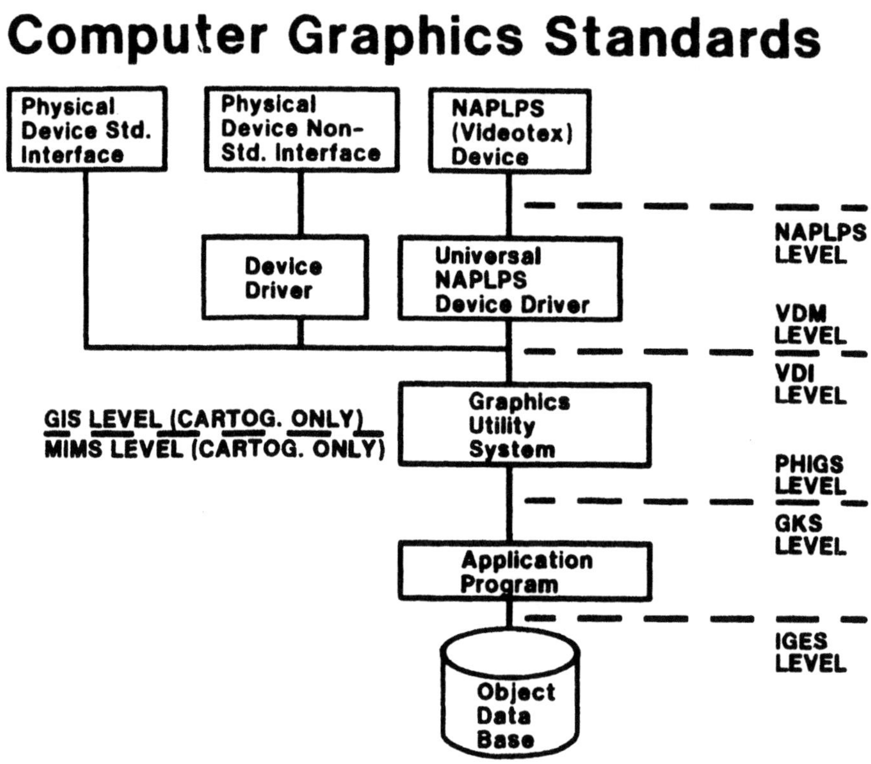

A brief description of the various standards follows:

- Initial Graphics Exchange Standard (IGES) is an ANSI standard for transporting Computer Aided Design (CAD) and Computer Aided Manufacturing (CAM) data within and between systems.
- Graphics Kernel System (GKS) is an official European standard adopted by ISO for defining two dimensional computer graphics and allows graphics software to operate independent of computer type or the graphics input/ouput devices. It provides the needed functions for a majority of applications with a minimum number of commands. ANSI has endorsed adoption of GKS as the American standard.

- Programmers Hierarchial Interface to Graphics (PHIGS) is a developing standard, similar in concept to GKS, for dynamic systems. PHIGS is currently under review by ISIS and has been proposed for acceptance for ANSI and ISO.

- Computer Graphics Interface (CGI) is a two-way protocol which takes place at the lowest level of device independence.
- Computer Graphics Metafile (CGM) is a standard file format for two dimensional graphics data.
- North American Presentation Level Protocol Syntax (NAPLPS) defines the method of presentation and storage for videotex graphics.
- Geographics Information Systems (GIS) is a developmental standard for capturing, editing and displaying geographic information for polygonal analysis and thematic mapping.
- Mapping Information Management System (MIMS) is a methodology for cartographic data base management.

Implementation of the above standards will allow the adaption of a wide variety of graphics software to new generations of processors. However, the portability issue can be addressed in part by developing the proposed cartographics system utilizing UNIX System 5.2 as the operating system of choice. Portability is one area where UNIX excels because it can be implemented on a wide range of processors from many manufacturers, such as the AT&T 3B20 processor and the DEC VAX 11/780.

UNIX has been the operating system of choice for many engineering and technical applications. And since it is the operating system of choice for the AT&T product and many of the new computers, we should use our own operating system. The AT&T Interactive Planning Work Station (IPWS), which is a large scale mapping system, was developed as a UNIX system and many of the techniques learned and developed with IPWS can be used to develop a detailed cartography system. One of the advanced techniques for programming mapping systems, such as CAGES, is the use of the high level language, GRADIAL, which is UNIX based.

Data Base Structure

The importance of the data base structure grows exponentially with the size of the cartographic application which demands a data base structure that can effectively handle large volumes of information and can link to other systems.

The information stored in the cartographic data base must be stored as a continuous map to guarantee minimal plotting error. Plot information is stored in latitudes and longitudes as x,y coordinates. The data structured of the continuous map is partitioned into user-defined geographic polygons that are referred to as facets. Each facet is unique because of the associated x,y coordinates which are used to integrate the facet into the continuous map. Facet size can be varied to coincide to data densities. In extremely dense areas, such as a center city area, a facet may represent only two to five thousand square yards, which may equate to several city blocks. Conversely, in less dense areas, such as the Mojave, a facet may cover many square miles. Using this level of data discrimination, a SMSA or Theodolite area may be defined by one or one hundred facets. This variable data sizing permits optimization of cartographic data.

By defining the continuous map of the U.S. as facets, the user can retrieve information not only by searching for a specific attribute, but also by directing CAGES to specific x,y coordinates, latitude and longitude, V and H coordinates or approximate graphic location. By approximate graphic location, the data base design can home in on site, county, city, municipal township, lakes and infrastructures.

Using the facet definition, the system response speed remains efficient as the number of graphics polygons grow because only a limited portion of the continuous map needs to be searched for any given display.

All the data required to describe the infrastructure, street numbering and municipal or company imposed boundaries, such as LATA's, are confined to a facet and are stored separately within the data base as specific non-graphics attributes and graphics icons (graphical representation of an element). In addition, linkages to other non-graphics data bases, such as the Planning Customer Data Base (PCDB), provide information such as specific customer address and location code, in V and H coordinates.

Separate data base files will be maintained for public domain data, such as the U.S. Census Bureau's GBF/DIME files of each U. S. Standard Metropolitan Statistical Area (SMSA). This will also hold true for files such as the Electronic Theodolite Survey data.

Because of the expected size of the AT&T-C cartography system complete data base management is required for both graphics and non-graphics data. The data base management capabilities provide the means by which information is stored and organized and, more importantly, the way that the multitude of facet data, map layers, icons and other data types are manged and interconnected.

The system manager schema organizes the cartographic geo-coded data in a unique set of tables and directories that provide the flexibility to be updated or modified to meet the demands of a dynamic interactive system. By utilizing the data base schema approach, a schema conversion program can be run by the data base administrator to incorporate new concepts and new pointer programs into the directory. Any number of tables can be defined and these can easily be expanded or changed at any time.

Applications Supporting the Development of CAGES

Many of the AT&T Communications organizations have shown an interest in having Systems Engineering pursue the development of a corporate cartography geo-code system enhanced with analytical capabilities. The application areas that were identified for inclusion in the CAGES feasibility study are listed below in alphabetical order. As part of the feasibility study, an economic value will be determined for each application and they will be rank-ordered by the value to AT&T-C.

- Access Planning
- Alternative Access Studies
- Cable Route Mapping
- Chorapleth Mapping
- Customer Locations
- Direct Service Planning
- Feature Groups for Local Service Offices (LSO's)
- Indigenous Customer Grouping Studies
- Local Access Transport Area (LATA) Mapping
- NPA and NXX Boundaries and Studies
- Outside Plant Protection and Restoration
- Point-of-Presence (POP) Planning
- Political Boundaries and Jurisdictions
- Population Density Studies
- Rate Band Studies
- Real Estate Records
- Scattergram Shadowing
- Site Location Topos
- SOMS FID Mapping
- Switching Studies
- Tax Maps
- Terrain Mapping
- WATS 800 Studies
- Zip Codes - 5- and 9- Digit Maps

Recommendation

The computer graphics environment is changing rapidly. Two years ago, there were approximately 60 to 65 vendors for both hardware and software. Today, that number is somewhere around 300 and this does not include service bureaus that specialize in engineering CAD and cartography offerings.

Along with the rapid vendor growth, processor capabilities and terminal and peripheral hardware features are increasing constantly.

Although there are many vendors offering "expert" turnkey systems, to utilize this method of satisfying the business requirements would create an atmosphere of vendor dependency. AT&T-C would be dependent on vendor for a specific processor, work station and terminal features and software releases. The latter may be the most critical because AT&T-C would not own the cartographic source code and would be dependent on the vendor for even minor system enhancements. And since an enhancement may be specifically for AT&T-C, we would incur special development costs.

To meet this changing environment, CAGES should be designed using an open architecture approach, utilize the various national and international graphics standards for maximum interface capabilities and be developed for the UNIX environment.

This development approach will allow AT&T-C to take advantage of public domain files and software, and allow the flexibility of incorporating specific vendor software, provided they adhere to the standards. But most importantly, the CAGES programs will be portable within the UNIX environment. CAGES will not be designed for a specific processor, nor will it be locked to a current work station design.

CAGES will use a layer structure for the system architecture. This will allow common use of the graphics GKS-type core and public domain layers by the entire user community. However, at the same time, CAGES can be partitioned to allow the creation of company proprietary layers, application data layers and private user data layers.

An important design feature of CAGES that is not easily attainable from turnkey systems is the inclusion of pointers within the various tables to allow CAGES to interface with other corporate systems and data bases, such as the Planning Customer Data Base (PCDB), and to retrieve required information. This ensures CAGES ability to stay syncronized with the latest data available.

The following figure represents the CAGES layered architecture.

CArtographic GEo-Coding System
CAGES
(Proposed Layer Structure)

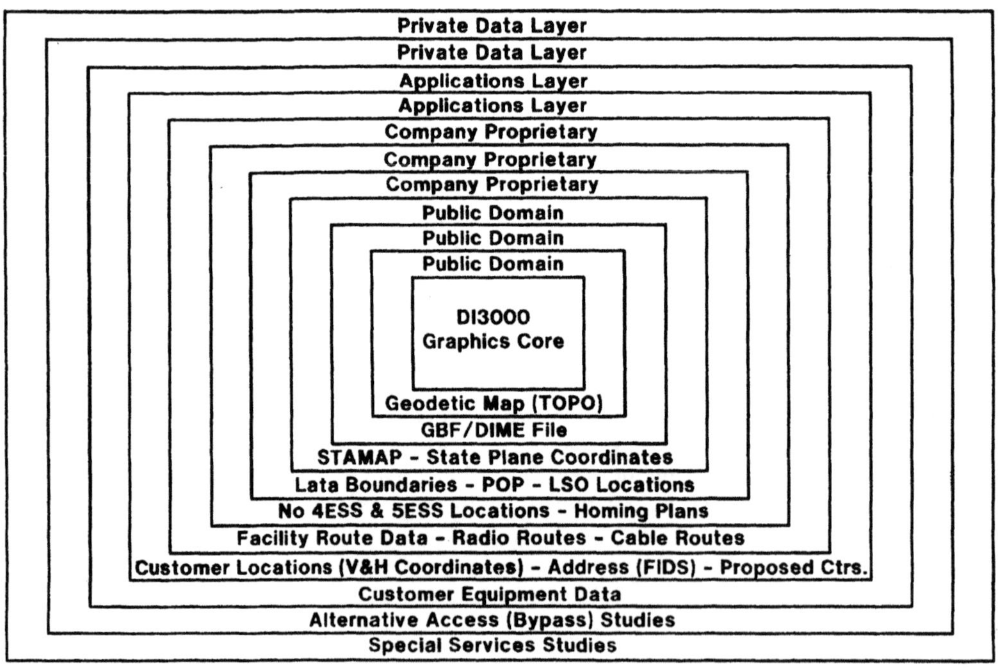

Biography

Mr. Bannon is the Manager - Systems Engineering Computer Graphics Consultant for AT&T Communications. He is responsible for planning and implementation of plans that direct the integration of Computer Graphics into the Engineering EDP environment. These plans incorporate new technology and concepts into the corporate decision analysis support process. He is responsible for all phases of computer graphics - CAE, CAD, CAM and Business Graphics. This includes earth science cartography systems.

Mr. Bannon studied for a BSEE at Pennsylvania State University and is a member of the NCGA, IEEE, IEEE Computer Society, and SAME. He is a member of the Advisory and Program Committees for CADCON and ATI. He has participated both nationally and internationally in conferences such as CAMP, MICAD and CG Tokyo and he has authored over 20 computer graphics papers.

6

CAD, CAM AND GIS: TOOLS FOR FACILITIES MANAGEMENT AND PLANNING

William G. Brodie
The Brodie Group
P.O. Box 55668
Seattle, Washington 98155 U.S.A.

Abstract

CAD, CAM and GIS systems are explained as applied to municipal and military engineering and planning needs. As the graphics and data processing capabilities of these systems are compared, definitions of the terms emerge. A plan for choosing a system and implementing a program is outlined. Computer-based graphics equipment now on the market provide planners, engineers and facility managers with well-developed tools; success with this technology depends on finding and using the right tools for the job.

1. INTRODUCTION

The management of permanent facilities, whether utility networks, building complexes or transportation facilities, requires extensive information about the condition and use of those facilities. The information is required not only for maintenance and operation of the facility, but is required for the planning and design of new facilities. With large facilities and small staffs, gathering and maintaining accurate information about the facilities has become a problem in itself. As-built records rapidly fall out of date, creating a credibility problem not just with a particular document, but with the entire records system. As a result, planners and engineers must spend considerable time verifying the accuracy of the records before they can proceed with their projects. In the extreme case, this verification means a costly and time-consuming field inspection and inventory of existing conditions.

In most cases, the facility engineer responsible for maintenance and operation is also called upon to maintain these records. Military bases have these responsibilities clearly laid out in regulation. For local government, the engineering department often has this duty. The records fall into two categories: graphic information (maps and drawings) and non-graphic information (card files, computer files, tabular reports). Ideally, a standard map set of the entire area of interest exists, complemented by large-scale site drawings that provide additional detail as needed. Up-to-date status information regarding conditions is readily available in an accompanying tabular report. Any and all maintenance and construction activity is routinely recorded at both the detail and summary levels. This ideal is rarely achieved, requiring extraordinary organization and considerable staff time.

Recent developments in computer technology have provided the facility manager with the opportunity to consolidate the graphic and non-graphic records into a unified record-keeping systems that has the potential to satisfy both the maintenance and operation needs and the long range planning needs. It is now possible to use automated systems to manage the information required for operating our public, private and military facilities. Full-scale automation is a large undertaking, however, and the costs can be staggering. The paper will explore the need for automation of these records systems and show how computer graphics and related computer developments are being used to address this need.

2. FACILITY MANAGEMENT AND PLANNING

Good management of our infrastructure systems results in good service to our constituents, whether they be customers, taxpayers, tenants or residents. Good service means efficient maintenance and effective planning for changing demands. Maintenance of the facilities includes emergency maintenance - responding when something extraordinary happens, and ongoing preventive maintenance. The planning function, on the other hand, develops strategies for responding to change. Change often means growth, where demands for new and expanded services (i.e. new facilities) arise. Without adequate information of the conditions as they exist, effective emergency response, preventive maintenance and planning activities are impossible. Facility records have only one purpose: to provide decision-makers at every level with the information they need to do the job at hand. Well-informed decisions are better decisions, resulting in more efficient use of all resources: less time and money spent delivering the required services. This is good management, this is good government.

Large facilities of any kind, whether public or private, face similar problems. The need for utility and transportation services is the same for military personnel living and working on post as it for civilians commuting from the suburbs to the factory. Public airport and seaport facilities, university and other public campuses face the same situation regarding space planning, building use and maintenance. The remarks that follow are based on the author's recent experiences with municipal and military engineering and planning functions related to utility, roads and property management.

MILITARY INSTALLATIONS

Although military bases resemble small cities in may ways, the very nature of their business sets them apart. Their utilities, roadways and buildings are not much different than those in any city, but because each base has a part in national defense, it can be dramatically transformed overnight and must always be prepared to make that transformation. Peacetime changes in personnel and facilities happen with reasonable lead time for planning and construction. Military emergencies, however, often provide little advance notice, either in scope or direction. Possible changes each base could undergo depend on its mission: its personnel and equipment could be called away or it could have to house thousands more. This need for sudden mobilization makes the need for facility planning critical within our Armed Forces.

Responsibility for facility planning within the military is well-defined, spelled out in regulations. Because the post master planning activity falls within the same command structure as the engineering, construction and maintenance activities, the opportunity to effectively maintain as-built records is as good as it is anywhere. Unfortunately, military planners often rely on records developed and maintained apart from the maintenance and operations people. Not only does this cost more because of the redundant information gathering, but it also creates problems when the information is found to conflict. Which is correct? Which should be used to gain budget approval for a particular project? Reconciling two sets of records adds additional time and effort to any project. Two sets of half-reliable as-built records is not equivalent to one set of reliable records.

Part of the difficulty is due to the fact that so many diverse activities can change as-built conditions. Even in the relative quiet of peacetime, maintenance, construction and demolition activities can be hectic. It is often not humanly possible to keep a centralized records up to date. This activity is frequently deferred until 'things slow down'. Rarely, it seems, have things been slow enough to fully consolidate facility information into a reliable, maintainable set of records.

Fiction, of course: Imagine the planner who must decide how to accommodate a demand for new storage space for automotive parts. The planner, new to the post, dutifully analyzes the requirements and, after considerable review of the maps and records, identifies a new building location. An available site was found on a road that would require only minor extensions to water, sewer, electrical and phone systems. The ground would call for minimal excavation. Satisfied with both himself and his preliminary siting, he presents the plan to his superior...

"Don't you realize the horseshoe pits are at this site?" The chief engineer evidently was not as satisfied, pointing out that a personal inspection of that site would have saved him time. The chosen site had been unofficially converted into the base horseshoe pit. Pits, actually. There were twenty pits, altogether. The first pit, originally just dug in the grass a couple of years ago, seemed to have been so popular that it multiplied virtually overnight. The concrete and sand for the pits and the slab for the basketball hoop were added when they poured the foundation for the maintenance yards, across the street. The site had evolved into full recreational use.

Still needing storage space for the auto parts, the embarrassed planner was directed to look for space in an existing facility since several large buildings were vacant. Eager to please, the young engineer found an appropriate vacant building, already serviced by adequate utilities. The project went ahead without a hitch, the need was met without extensive new construction. It was about 18 months later when the property manager realized that the 40-year old building, scheduled for destruction, would have to be regarded as a 'temporary' structure for perhaps another decade. End of Story.

Accurate and available. Whether in fiction or in fact, facility use changes almost daily. Decisions regarding building use, maintenance and new construction are made just as frequently. For the best decisions, it is clear that the same information about changes must be readily available to all the players. It's not sufficient that an accurate as-built drawing exists, it must also be available to the property manager as well as the planner. It's not enough that long-range planners do an exhaustive inventory of building use for their reports; their results must be made available to all.

Military bases, as small cities, face the same ongoing maintenance, operation and planning issues as do non-military facilities. The difficulties of as-built record keeping are universal, adding cost and confusion to planning efforts. Solving this difficulty will require not only a means to keep a set of accurate, up-to-date records, but they must be readily accessible by engineers, planners and operations staff as they make decisions. Good decisions depend on good information.

MUNICIPAL FACILITIES

Local governments, unlike military bases, are not called upon for full military mobilization. The local engineers, usually within a Public Works department, have plenty of complications to more than adequately equalize the headaches. As we have stated, the infrastructure systems that provide utility and transportation services are, for the most part, just like those on military bases. The need for building and property management is equally, if not more, important at the local level. The political and administrative environment in many municipalities creates management and record-keeping requirements not often experienced in the military. With a hundred sources of funds and a thousand 'watchdog' citizen groups, the local government is the focal point for many different forces in our society.

For the typical city or county Public Works department, successful performance in the streets means political success. Taxpayers and ratepayers watch carefully and are much less apt to forgive than the enlisted men. The public is not the only body making demands, the Planning department, the Tax Assessor, the neighboring city, the private gas company...they all need to know the status of this project, the cost of that one, the depth of this water main, etc. The cry from the outside demanding information is just as loud as the internal one, often with a politically powerful voice.

Information, please: Delivering utility and transportation service to the American public is an immense task. Again, accurate as-built information is a basic requirement for effective maintenance, operation and planning. Underground facilities form mind-boggling webs, usually with multiple owners. Mistakes in design or digging can be extremely costly. Inaccurate or out-dated records result in just that kind of mistake. Despite noble efforts to coordinate digging activity, we find examples of horror stories on a regular basis. If we can't keep our internal records up to date, how can we hope the give the underground locating coordinators good information?

This applies for both overhead and underground utilities as well road systems. Intelligent infrastructure maintenance management might be possible if we really knew what all the existing conditions were and what other improvement and maintenance activities were being contemplated in the area. How often has a fresh resurfacing job been scarred with a scheduled utility dig? These problems are not new ones, but our success in dealing with them has fallen short of the goal. The cost of foul-ups is too great to give up the fight.

We'd have a better chance at managing our construction and maintenance programs if we could only manage the information. Fortunately, facilities data management (i.e. gathering, storing and communicating this facility information) has become easier with the advent of the computer revolution.

3. CAD, CAM and GIS DEFINED

Before the application of computers to these facility record-keeping problems is discussed, the terms will be defined. The definitions are quite narrow and are not universally accepted, but will help to classify systems for us here. The terms CAD, CAM and GIS in this context will refer to specific classes of computer capabilities and their use in geographic data processing. It's important that the differences be understood; a key conclusion of this paper depends on this understanding.

COMPUTER-AIDED DRAFTING

"CAD", as it's used here, is the use of a computer to compile, modify and produce a drawing. In its simplest sense, the CAD operator has replaced his drawing tools and table with an electronic substitute. The drawings reside inside the computer memory and can be developed and later modified electronically. The drawing can be developed and reviewed on a graphics terminal and later transferred to paper or mylar via an output device, such as a pen plotter. The capabilities of CAD systems vary in their ability to perform certain drawing functions automatically, complete systems begin under $20,000.

Confusion about CAD systems is partly due to the use of the abbreviation for Computer Aided Design. CAD/CAM as it was originally and is most widely used, refers to the Design and Manufacturing process, not drafting and mapping. In our context, CAD refers to computer-assisted 2-dimensional drafting, tools for the production of maps and design drawings.

COMPUTER-AIDED MAPPING

"CAM" would be better called mapping information systems (however, we can't use MIS as an acronym - it's also been taken). As used here, CAM systems often involve the very same computer equipment as used in CAD systems but are to be distinguished from CAD both by the intended products (maps) and certain software capabilities. The first difference is simply that we're using CAD technology to draw maps rather than wing assembly diagrams or wiring schematic drawings. In a similar way we can call the family station wagon a 'bus' when we haul the kids to baseball practice, and a 'delivery truck' when we fill it with sheets of plywood for our next home project. Specialized use of CAD technology for maps hardly justifies another acronym, but the second distinction is an important one to grasp.

System capabilities for handling map-related textual information and for storing very large geographic areas distinguish CAM systems from CAD as defined above. Every map feature, such as a building outline, man-hole, contour line, etc. has a story to tell that goes beyond the labels on a map. It may be a summary of building use, a scheduled maintenance date or an indication of what the source data was. With the latest CAM (map information) technology, this related information can be stored right along with the graphic element in the computer, even though the information never actually appears on a map. Inquiries based upon this behind-the-map information can produce specialized maps and reports that isolate only those features which pass the test. For instance, a 1" = 1000' scale map showing only water mains installed prior to 1950 can be easily created, provided the installation date was entered as the water mains were put in. Also, this non-graphic information can be tabulated by the computer into any zone or region that can be drawn. For example, total vacant building space for a particular study area can be tabulated simply by pointing out the area in the computer.

A second major distinction between CAD systems and CAM systems is the ability to store and manipulate very large geographic data sets. Usually, this means a large geographic area that comprises the entire mapping base. It may mean an entire airport facility, Army base, city, state or even a country. Few CAD systems have effectively dealt with this issue. Theoretically, they can process a 'drawing' of any size and complexity. In reality, most CAD systems suffer dramatically as the amount of drawing points gets larger. Mapping systems have the ability to store and manipulate these large data sets without significant loss of speed. These advanced capabilities set CAM systems apart from the single-sheet oriented CAD systems.

Due to their extensive data management capabilities, CAM systems often require large super-mini computers and special graphics devices. Their costs are easily over $100,000 and approach $1 million in large installations.

GEOGRAPHIC INFORMATION SYSTEMS

"GIS" traditionally refers to systems that have the ability to analyze problems, given the required geographic information. This type of system is not as easily characterized because a wide variety of problems that have a geographic component are solved with the help of computers. This class of system resembles CAM in that it stores information, but it goes beyond the basic record-keeping functions that characterize CAM systems. GIS systems have the ability to model and analyze geographically oriented problems. The information structures in these systems typically allow for network, polygon and grid-oriented information schemes. Because the data are carefully connected by the computer, a number of problem-solving processes can take place.

The ability to logically relate one 'layer' of map information to another is one of the possibilities. For instance, a GIS system could relate soils information with land use information and determine the most likely site for an environmentally sensitive building project. In these systems, networks are not only graphically depicted, but are logically connected so that computerized network analysis can be accomplished directly out of the mapping data base. This is a common requirement for capacity planning in utility and transportation networks.

GIS software also may require large super-mini computers because of the heavy computational requirements. A packaged system with extensive GIS tools will exceed $100,00 and continue up.

To summarize, three classes of computer systems useful in facilities management are identified along with their characteristic capabilities :

> CAD systems are electronic drawing systems capable of automating a draftsman's tasks, building a multi-layer drawing with no additional associated information.
>
> CAM systems extend CAD capabilities, providing associated attribute information to the graphic elements on each layer of the map. Also, large geographic areas that include many map sheets can be efficiently processed.
>
> GIS systems logically connect the geographic data to allow for computer analysis and problem solving. They don't necessarily have high-quality, "draftsman" look to their map products like a CAD or CAM system. The primary purpose of the output maps is to display the results of the analysis, rather than a finished map product.

Commercially available systems will rarely fall neatly into any of these categories. Leading CAD systems often have modeling and 3-D capabilities that go far beyond the basic capabilities described above; some GIS systems have well-developed output graphics, others do not. The reason for the distinction is to highlight the different capabilities and dispel the notion that CAD, CAM and GIS mapping systems are all pretty much alike. They are not, and the well-informed user of these systems understands the differences before he begins any significant project.

4. COMPUTER-AIDED MAPPING PROJECTS

CAD, CAM and GIS tools can all help engineering, maintenance and planning people make better decisions related to facilities. The key to successful use of this technology is to embark on the project with a clear understanding of a) the needs and problems at your facilities and, b) the available solutions. Classifying your needs as either CAD, CAM or GIS may help in sorting out the confusing array of products.

Increasingly, engineers and planners around the world are recognizing the possibility to automate all their facilities and plant records into a comprehensive computer-aided mapping and information system (CAM). These programs also often incorporate CAD (drafting) and GIS (analytical) capabilities. Not only are the maps and drawings being compiled into the computer, but all the card files and computer files with related information are being brought in, too. For facilities of any size, this turns out to be a significant task, with budgets often measured in million of dollars. Four distinct work areas emerge in the successful automated mapping projects: study, design, construction and implementation.

STUDY, DECIDE

The purpose of the study phase is to define the objectives and scope of the project and to provide management with sufficient cost and benefit information to decide how to proceed. Depending on the scope of the project, This phase may extend over several months (even years) due to the long-range corporate commitment it is seeking. Some of this analysis is done informally, but with projects of significant size, the following steps should be followed:

<u>Identify the areas of greatest need and opportunity</u> for automation. Drafting backlogs, redundant and inconsistent map sets, lack of reliable maintenance records, etc. are among the areas to check.

<u>Propose alternatives</u> for addressing the problems. A range of alternatives should be presented and analyzed for costs and benefits. A isolated CAD approach may have some immediate benefits. The costs and advantages of using professional help for project design and map digitizing should be examined.

<u>Obtain management commitment.</u> At this point, a decision on the alternatives should be made. Resources for the project should be committed and the design work begun.

DESIGN, DECIDE

The objectives of the design phase are to specify in the products, services and capabilities of the new system. As this is done, the budget for the project is refined. Requirements for automated drafting capability (CAD), geographic analysis (GIS), and facility map information processing (CAM) are determined. Having identified the goal, the path is then laid out: staff, procedures, services and computers are specified.

<u>Document specific requirements</u> for the new system. This is where a clear understanding of CAD, CAM and GIS capabilities is needed. The content, scale and accuracy requirement for every desired map product is determined. Requirements for analytic or data processing functions are spelled out. This step describes the exact nature of

the new computer facility; everyone involved should review this description. It tells them what they're buying.

Design the procedures. At this stage, the plans are specified for new field data gathering, preparation of existing records, map digitizing procedures, computer processing and related services. As a part of this grand design, a refined cost estimate is made. Often, this calls for a pilot or prototype effort to refine the design.

Specify the computer system. Related to the item above, a technical specification for a computer system is prepared. Most often, the required functional capabilities are presented to the vendors for competitive bids.

CONSTRUCTION

The objectives of the construction phase are to build the computer data sets required for the project. The land and facilities data will form the information base from which the CAD, CAM and GIS functions will proceed. This phase is the most expensive because of the volume of data being copied to digital format. Because of the special equipment and skills required, some data base construction steps are more efficiently done by service companies specializing in this. This phase may go on simultaneously with the implementation phases.

Gather required field data. New field surveys, aerial photography and condition inventories may be required, depending on the quality of existing information.

Prepare existing records. Map records should be brought up to date. Conflicts between records should be resolved before they get into the computer system.

Digitize the records. This would include both the map records and any related non-graphic facility information that is to be a part of the system.

IMPLEMENTATION

Implementation is the bringing of all pieces together to begin delivering the map products, CAD or GIS services to the engineering and planning community. Successful implementation will require readiness on three fronts: the computer system, the personnel, and the data base. Each of these three elements must be kept at their best for continued success.

Install the computer system. The system includes all the software and equipment. For large systems, planning for space, power, climate control and data communication facilities should begin early in the process.

Train personnel. Management, professional and technical staff will all play a role in the use of the system. Technical training should include workstation skills, data base configuration and programming. Software vendors cannot provide all the training that is required.

Managerial and professional staff require orientation that relates to their own data and analytical requirements. An in-house program is called for.

Install the data base. The digitized map information may be coming from another computer system or perhaps was developed on the same system. It's installation also implies that procedures are in place to keep the data base up to date. The records management of the organization should have re-arranged itself to feed into the new system.

Projects of this nature distinguish themselves due to the multi-disciplinary requirements. Professional disciplines of data processing, civil engineering, land surveying, photogrammetric engineering, cartography, etc. all may have a role at some point in the project. Management of the conversion efforts calls for a good understanding of the nature of project, having elements similar to traditional data processing projects as well as construction projects.

5. SUMMARY

The information required to intelligently manage large facilities is staggering in both volume and complexity. Maintenance, operations and planning personnel require the much of the same map and as-built information, but often rely on different and inconsistent records systems. The unreliable maps and records result in decisions made on hunches or poor information. The results can be expensive due to additional field work and office verification of records. Worse, maintenance management and capital improvement planning can result misguided spending.

Computer graphics and data base technology is currently available to address these needs. Computer-aided drafting (CAD) can go a long way towards making the drafting function more efficient. So-called GIS systems can establish computer models of geographic problems and analyze different sets of requirements. CAM systems can store the detailed map records along with the related facility information and provide the tools for a comprehensive facilities records system.

A comprehensive, automated facilities records system is entirely feasible with appropriate project management and design. Such programs are underway in several cities, utilities and military bases throughout the world. These projects involve much more than installing a new computer graphics system, and require extensive planning. Once operational, planners, engineers, maintenance and operations staff can all be served by consistent, accurate maps and related information. The result is better management, better government.

BIOGRAPHY

Mr. Brodie is director and principal consultant of The BRODIE Group, a Seattle firm specializing in the design and management of computer-aided mapping and geographic information programs. He was the system consultant and project manager for the City of Bellevue, Washington's successful computer mapping and GIS program. The firm has consulted with local governments, utilities, school districts and federal agencies providing design and management services for geographic information projects. A graduate of the University of Washington in Seattle, Mr. Brodie has worked in the computing field since 1972.

7

MANAGING A DISTRIBUTED CAD/CAM SYSTEM

Dale O. Christensen
Naval Sea Systems Command
Washington, DC

Bryan L. Hill
Naval Weapons Center
China Lake, CA

ABSTRACT

The management of the largest distributed CAD/CAM system in the United States is the responsibility of the Naval Weapons Center, China Lake, CA. This paper describes the acquisition strategy, implementation and operation of this distributed system.

INTRODUCTION

The management of distributed system presents a tremendous challenge to any would-be manager. Add to this basic problem the elements of a rapidly evolving technology, the necessity to cross both organizational and functional boundries and the requirement for public accountability of all management actions and the challenge of successfully managing the United States Navy's computer aided engineering and documentation system becomes truly formidible.

The United States Navy's Computer Aided Engineering and Documentation System (CAEDOS) is currently installed at sixty-four separate Navy sites within the United States. It represents the single largest centrally acquired and managed distributed Computer Aided Design/Computer Aided Manufacturing (CAD/CAM) system in the world today. CAEDOS equipment is installed in all Navy laboratories, shipyards, aircraft overhaul facilities and public works centers. Local on-site communications for CAEDOS is provided by a star network while a 4800 baud modem allows data file transfer between sites. Currently, limited amounts of data are already being transferred among sites.

Successful integration of a CAD/CAM system in a distributed environment is, at best, a difficult and tedious task. The success of CAEDOS can be attributed in large measure to the process of goal setting and extremely detailed planning which preceded system acquisition and installation. The goal setting process for CAEDOS involved Navy management, the Navy user organizations and the personnel tasked with system acquisition, installation and day-to-day management. If an organization is to be successful in integrating CAD/CAM, it must initiate an analysis and goal-setting process which provides long-term direction for the program. The basic theme of this paper is that successful management of a distributed CAD/CAM system must start long before system installation. Basic goal setting and long range incremental planning are essential to optimum system utilization, integration and management.

The management of a distributed system, such as CAEDOS, wherein the user sites are not organizationally linked to the CAEDOS program manager, as would be found in a centralized or hierarchical organizational structure, requires that other methods of enforcing system standards and discipline be utilized. In the CAEDOS environment that discipline has been provided by centralizing contract administration and contract payment within the CAEDOS program office.

The seven-year history of CAEDOS accomplishments results primarily from the timely definition of goals and the incremental effort to accomplish those goals. No one single item led to the successful acquisition and integration of CAEDOS. Rather, it was the result of four major phased efforts. The initial phase started in October, 1977, and included a one-year study to define the CAD/CAM needs of the Navy laboratories and to identify the most promising solution for those needs. The second phase, the acquisition phase, lasted three years and was used for specification development and contract execution. The third phase was dedicated to system installation, inspection, acceptance and initial user training. The fourth phase is the day-to-day sustaining management of a productive distributed CAD/CAM system. This paper will attempt to highlight the basic goals of the CAEDOS program, the phased implementation of those goals and the manner in which that process contributed to successful management of the distributed CAEDOS system.

Initially a CAEDOS program manager was appointed by the Naval Material Command, parent organization of the Naval Laboratories, to pursue a requirements definition study. Each Navy Laboratory appointed a point of contact for user input to that study. Ultimately, that effort produced five basic goals. Those goals (which became the basic long-term directions for the CAEDOS program) were:

(1) CAEDOS should represent a single, general purpose, standardized CAD/CAM system which could be configured to meet the specialized needs of the various Navy Laboratories. The acquisition of the CAEDOS would be fully competitive and all eligible system manufacturers would be encouraged to bid.

(2) A central design agency (CDA) should be established to perform initial system inspection and acceptance (benchmarking, specification conformance, hardware and software testing) and to assist each user site with system installation and acceptance. The CDA would be

responsible for developing standard menus, libraries, naming conventions and other appropriate standards, for test and validation of the system manufacturer's software releases, and for developing applications software for use at multiple sites.

(3) Initiation of a productivity study which would assess the impact of the CAEDOS on the Navy user organizations.

(4) Each laboratory would write an implementation plan for its CAEDOS which would be reviewed by the Naval Material Command.

(5) The Navy would support the development of the Initial Graphics Exchange Specification (IGES), a neutral file format to facilitate the exchange of graphics data between unlike CAD/CAM systems.

These five relatively simple goals formed the basis for the second and subsequent phases of the CAEDOS implementation effort.

The second program phase was the acquisition phase. In March, 1979, the Naval Weapons Center, China Lake, CA was selected to be the Navy's lead laboratory for CAD/CAM. The CAEDOS program manager established two separate functional groups to support the second phase effort. The first group, entitled the CAEDOS program office, was tasked with the acquisition of the CAEDOS while the second group, the CAEDOS Central Design Agency, was to become the Navy Laboratories center of expertise in the installation, operation and support of CAEDOS.

The acquisition phase consisted of the specification development and competitive acquisition of the CAEDOS CAD/CAM system. The major goals of the acquisition phase were to acquire an affordable, state-of-the-art, commercially available, reliable, CAD/CAM system meeting all Navy requirements. Under the CAEDOS contract, the system manufacturer was required to provide continuing training, documentation, hardware maintenance, software maintenance, and miscellaneous engineering support.

Since the Navy Laboratories formed the basic user group for the CAEDOS, the acquisition of a true state-of-the-art system was of prime importance. The CAEDOS contract specification had to reflect the sate-of-the-art nature of the user's requirements. The user's requirements and the contract itself in turn had to provide a mechanism for system technical upgrade as necessary.

The CAEDOS program office utilized the specification development process to initiate a dialog with the Navy's targeted user organizations, a spectrum of large industrial users, and also with all interested CAD/CAM manufacturers. The philosophy behind this extensive consultation effort was to more effectively describe the basic system requirements in the context of CAEDOS program goals. An ad hoc group of prominent industrial CAD/CAM systems users including representatives of the Boeing Commercial Airplane Company, General Motors' Saginaw Steering Division, the General Electric Corporate Research and Development Center, the Xerox Corporation, Sandia Laboratories, RCA, Weyerhauser and Brigham Young University (a leader in undergraduate CAD/CAM education) reviewed the specification as it evolved and provided valuable insight concerning both its strengths and weaknesses. In parallel with the development of the specification from a users viewpoint, the major manufacturers of CAD/CAM systems were contacted and provided with the opportunity to comment on the evolving specification. Each interested manufacturer was encouraged to provide those comments to CAEDOS program management in face-to-face, day-long individual meetings. The result of these user and manufacturer discussions was the development of a specification which provided a true state-of-the-art solution to the Navy's user requirements.

A number of basic system concepts helped shape the CAEDOS specification. Since, the Navy realized the necessity to acquire a true state-of-the-art system which incorporated advances in the system state-of-the-art, some system capabilities were included in the contract as delayed compliance items. The contract also allowed for the purchase of any new beneficial software that became available during the life of the contract. Upward compatibility of the system software and data base(s) was required and was obviously critical to long-term system success.

CAEDOS program management required CAEDOS to be in the mainstream of the commercial development process. The Navy did not want and could not afford an "orphan" or "Navy unique" system that the system manufacturer would lose interest in supporting soon after system acceptance and payment. Another key element was the system availability level. How valuable and useful is a state-of-the-art, upward compatible system, one which is in the mainstream of the commercial development process if it is not available for use? A number of separate efforts contributed to mitigating that problem such as inspection of the manufacturers plant, review of his field service organization and incorporation of a stringent system availability algorithm into the contract.

Although not one of the original CAEDOS goals, it became obvious after discussions with both Navy and commercial users that it would be necessary to upgrade user training and documentation then available in the CAD/CAM market place to match the technical capability and complexity of CAEDOS. The CAEDOS contract required the contractor to rewrite all documentation to Navy standards. This requirement ultimately provided the NAVY with much improved user training and documentation while also providing the manufacturer with an improved product to market to other potential CAD/CAM buyers. This improved documentation was therefore provided to the Navy at a very reasonable price which in turn contributed to the ultimate affordability of CAEDOS. Presently there are 35 different classes available under the CAEDOS contract. This includes classes for upper and middle management of user sites, for CAEDOS system managers, and for the spectrum of CAEDOS users.

Computer system acquisition in general and CAD/CAM system acquisition in particular continues to be beset with major problems. Because of the continual advance in CAD/CAM technology, determining if a specific capability is available on a system requires much more than a review of manufacturers' sales literature. The solution to this problem in the case of CAEDOS was to require the contractor to commit in writing to which capabilities his system could perform and then to perform a benchmark test to validate that capability. The philosophy that if you don't perform a benchmark test, you get what you deserve has been proven over and over again in the CAD/CAM industry.

During the CAEDOS acquisition cycle, a "work throughput demonstration" concept was utilized. The purpose of this demonstration was to link technical capability to the CAEDOS system total life-cycle cost. The application of this concept provided a very effective methodology for CAEDOS acquisition wherein the most productive system was acquired for the Navy user environment.

During the Navy's internal Automatic Data Processing review and approval process, additional systems needs were identified which fell outside of the basic quantity of Navy Laboratory systems. It was therefore determined that an indefinite quantity type contract utilizing delivery order provisions would be the most appropriate vehicle for allowing the incremental, phased introduction of this technology into the Navy environment. The CAEDOS contract was ultimately awarded to the Computervision Corporation of Bedford, MA. The initial estimated contract dollar value was approximately $63,000,000. Within 12 months, the original contract was fully subscribed. The contract has been expanded to provide for additional work stations. The present estimated dollar value of the contract is approximately $99,000,000 over the eight-year contract term.

The first CAEDOS system was shipped to the central design agency at the Naval Weapons Center, China Lake, CA. There the system endured an exhaustive sixty day acceptance test. After the acceptance of the first system, the additional systems for the Navy Laboratories were installed on an incremental basis. The contract required CAEDOS system installations to be performed at three-week intervals with all Navy Laboratories to be provided with an initial system. Six months after system acceptance, the initial systems were enlarged on an incremental basis until the final configuration was reached.

The fourth and current phase of the CAEDOS program is the production phase. The first nine to twelve months were involved with initial system installation and user training. Beyond the system installation and user training phase, the productive use of the CAEDOS system begins in earnest.

The history of the CAEDOS program provides the basis for discussion of the successful management of a distributed CAD/CAM system. The success that has been seen to date is based on the extensive planning and implementation work that has been performed over the last seven years and on the content and controls of the CAEDOS contract.

One of the key elements in managing a distributed CAD/CAM system is providing a system which the users find desirable and usable. There are many features of the CAEDOS which are very attractive to the CAEDOS user. These features include the low initial acquisition cost, the state-of-the-art nature of the system, the reliability of the system and the required high level of system availability. The path to migrate from existing Navy CAD/CAM systems to CAEDOS and to and from other existing contractor's systems, the commonality of systems within a Navy activity and within a group of Navy activities, the ability to expand the system, and the ease with which equipment can be ordered are all attractive systems characteristics. All of these factors have brought a spirit of cooperation and friendship to the user community. A user who values this new technology will support the system because of the benefits to him. Distributed systems management is much easier when the system is valued by the users.

The CAEDOS contract structure provided users with an affordable, easily acquired system by maximizing competition, involving industrial users and producers in the specification formulation stages, and providing for optional system quantities. After the CAEDOS is installed, the users acquire a reliable and available system which is common with other internal and external Navy activities as well as many of the Navy's contractors. Requiring these attributes in the CAEDOS contract required many extra months of effort to include, but they were well worth it.

The positive acceptance of the CAEDOS has allowed for the users and managers to properly communicate needs to each other and do so in a positive environment. This is critical to management success. Integrating a new technology into an organization is difficult. If the managers and users concentrate on the deficiencies of the system rather than looking at the impact of the new system or technology on the organization, the major management issues will never be dealt with in a timely fashion. Proper management requires the major focus to be on the people and the organization, not the system.

The challenge of managing a distributed CAD/CAM system obviously is less demanding when the system hardware, software and documentation are identical. Central contract management and payment allowed the Navy to provide a centralized point of contact for system problem resolution with the system manufacturer and assures equal treatment of all user sites. The parochial interests of individual sites are minimized. This methodology has proven essential to successful system management. It allows the Navy to speak to its contractor with one voice. This is critical if priorities are to be set and followed.

The concepts of centralized trouble reporting and resolution, centralized hardware and software configuration management, and coordinated system enhancement are very important to good system management. All CAEDOS sites are linked today by a national communications network to the CAEDOS program office at the Naval Weapons Center, China Lake. A specially designed data base management system has been developed which provides on-line status of all CAEDOS sites throughout the continental United States and Hawaii. This system, the Problem and Change Management (PCMS) system, provides problem reporting/disposition, manufacturer field service monitoring, and hardware and software configuration management. This system, coupled with centralized contract administration and payment functions, allows the CAEDOS manager both the system status information necessary to make informed decisions and the authority to implement those decisions. Additionally, PCMS greatly enhances the ability of user site managers to quickly determine if system problems experienced at their site are unique or are common across like Navy systems. Manufacturer responsiveness to systems troubles, when they can be demonstrated to be the manufacturers rather than the users, is much enhanced by this centralized single line of communication represented by the CAEDOS program manager backed up by an on-line site status data base.

CAEDOS is just the first step in integrating CAD/CAM and Computer Aided Engineering technology into the Navy. A major benefit of the CAEDOS program is that line management has focused on CAEDOS. This organizational focus is a major step in the educational process of management. Most middle and upper management personnel are just learning about the integration problems of computer related technologies. This organizational focus is needed to adequately deal with the allocation of resources which management controls. Funding, space, and personnel issues must be dealt with effectively if any automation project is to be successful. Training and retaining personnel is a continuing effort due to the ever increasing demand for these highly skilled personnel in the marketplace.

The distributed CAEDOS system, now in its fourth year of existence, has provided the Navy with a contemporary, reliable, standardized, distributed CAD/CAM system. Centralized management of this system has provided the discipline that has allowed the Navy to take full advantage of these system attributes. Successful management of a distributed CAD/CAM system starts with the long-range, goal-setting process not with the acceptance of the installed system. Setting in place those necessary management tools such as PCMS, establishing clear lines of authority and responsibility for system management, developing a central design agency, enforcing standards and maintaining hardware/software configuration management are simply some of the actions that must be taken in parallel with the development of a system specification and acquisition of a system meeting those specifications.

CONCLUSION

The efforts of the Navy in implementing the CAEDOS system have been rewarded by the existence of a highly capable, productive and economical computer aided engineering environment within the Navy community. The benefits of this environment are only now being felt by the Navy's industrial base.

BIOGRAPHIES

Dale Oliver Christensen

Mr. Christensen is presently the Naval Sea Systems Command (NAVSEA) Division Director for Computer Aided Design/Computer Aided Manufacturing (CAD/CAM). In this role, he is planning the next generation of CAD/CAM systems to be acquired and developed by the Navy. His previous assignment was the program manager for the Computer Aided Engineering Program for the Naval Material Command Research and Development Centers. In this assignment he oversaw the procurement and integration of the Computer Aided Engineering and Documentation System (CAEDOS) at fifty Navy field activities. He serves on the Board of Directors of the National Computer Graphics Association (NCGA) and is a member of NCGA, Sigma Xi and the American Society of Mechanical Engineers. He holds B.S.M.E. and M.S.M.E. degrees from the University of Utah.

Bryan L. Hill

Bryan Hill is currently Head of the Information Systems Resources Division at the Naval Weapons Center. He holds a Bachelor's Degree in Economics and a Master's Degree in Research and Development (Public) Administration from the University of Southern California. He was a charter member of the AIAA's Technical Committee on software systems. Prior to Mr. Hill's participation in the Computer Aided Engineering and Documentation Systems Program, he was Head of Procurement and (later) Contract Management for the United States Navy's Research and Development efforts in tactical air launched weapons.

8

THE IMPACT OF GEOGRAPHY AND DISTRIBUTED DATA COLLECTION ON THE ENGINEERING OF ARMY TELEPHONE SYSTEMS WORLD-WIDE

Coy Correll
Senior Systems Analyst
Cochise, Incorporated
La Mesa, Ca.

Abstract

The United States Army Electronic Systems Engineering Installation Agency has the responsibility for developing and maintaining the Army's strategic communications systems. Part of this responsibility involves the upgrade of telephone systems on approximately four hundred posts world-wide.

This paper will discuss the role of automation in the engineering of the distributed digital telephone systems which will be used in this upgrade.

This paper will also discuss the impact of geographic information on the automated engineering system and the issues involved in current data collection efforts. Future engineering systems and remote data collection scenerios will also be discussed.

1. INTRODUCTION

In 1979 the Army began a program to upgrade its telephone system throughout the world. This effort is known as the Base Communications Program (BASCOP). BASCOP marked the beginning of a fundamental change in telephone engineering philosophy to take advantage of the latest telephone technology. This program marks the shift from mechanical to electronic telephone switches. This change in switch technology means greater reliability and a cost reduction for switching equipment. As a result, the outside cable plant becomes the most expensive part of the telephone network in terms of initial cost and maintenance throughout the life cycle of the telephone system. To provide greater reliability and lower the cost of the cable plant switches can be distributed throughout a post.

2. BASCOP

GENERAL

The distributed switching concept for engineering telephone systems created too many design alternatives for the engineer to analyze manually. Because of this, the Army decided to develop an automated engineering system to aid the engineer. The system uses rights-of-way for cable installation, locations for switches and telephone requirements for each building on a post to determine the most cost effective system installation. After a system design is selected by the engineer a bill-of-materials and installation drawings are produced.

SITE SURVEY

A site survey is conducted by an engineering team for each post to be upgraded. A typical site survey takes 3-6 weeks and 3-5 people. During that time three things are be done. This team reviews plant-in-place drawings to verify the current cable routes and identify other possible cable routes. Field surveys are conducted to identify pertinent geographic features and select buildings which can be used for new switch locations. Current telephone installation records and projections of telephone requirements for the next five years are reviewed and recorded. This information is brought back to the engineering facility for entry into the automation design system.

ENGINEERING

The location of all buildings and the cable right-of-ways are entered from the plant-in-place drawings. Telephone requirements are assigned to each building. The engineer then identifies locations for switches and the area of the post each switch will serve. The system then determines the lowest cost configuration of switches and cable plant which will satisfy the telephone requirements of the post. The engineer can then analyze this configuration and make any changes in switch locations or service areas. A new configuration can be produced based on these changes. When the telephone configuration is the lowest cost and meets the requirements set by the engineer an engineering installation package is produced.

INSTALLATION

Once a design has been chosen for the new telephone system, a set of drawings showing the cable routes, cable size and location of the switches is produced by the engineering design system. A bill of materials which specifies the quantities of cable by type, gauge and size; switch equipment and other miscellaneous equipment required for installation of the new system is produced.

3. GEOGRAPHY

GENERAL

The geography of a post has an impact on the automated design process in two ways. First, natural or man-made obstructions, such as roads or lakes, must be identified. Any obstructions which cross rights-of-way will impact the cost of cable installation. Second, the topography of the area must be examined. Because plant-in-place drawings do not show terrain features the true distance along rights-of-way in rugged areas must currently be manually calculated and entered. This allows the automated design system to correctly compute cable costs for routes which would be affected by any hilly terrain.

OBSTRUCTIONS

Currently, obstructions are identified by using plant-in-place drawings and field surveys to locate any obstructions not shown on the drawings. The distance the obstruction impedes the right-of-way is measured for use by the automated design system. Installation cost for portion of the cable is computed on a cost per foot basis. Currently, only one type of obstruction is recognized by the system. Future generations will recognize many types of obstructions.

TERRAIN PROFILES

The field survey team has one of two ways to deal with terrain which may impact distances shown on plant-in-place drawings. First, the distance may be measured. This is, in most cases, the easiest and most accurate. When it is not possible to measure the distance directly it may be approximated by using contour maps. This will normally provide the accuracy required.

4. DISTRIBUTED DATA COLLECTION

GENERAL

The collection of current and future telephone needs provides the information required to determine the proper size and cost of various routes. Currently this data is collected using transportable computers and brought back for validation and data entry. If there are errors or inconsistencies in the data, a team must be sent back to resolve them. This, of course, is expensive and time consuming. To resolve this problem, two alternatives are being explored. The first is an interm approach while the latter is would require a more gradual and expensive implementation.

PORTABLE SYSTEMS

The most direct solution to the problem of remote data collection is to take advantage of the rapid advances in technology and replace the current hardware with smaller, more powerful systems. This would allow newer, more sophisticated programs to be developed. This generation of the software would be capable of doing both the data collection and the validation in the field. After the data has been validated, it would be sent to the engineering facility via modem. This could become a long term solution, if the processing power and display capabilities of portable computers improve enough to make portable CAED systems feasible.

POST FACILITIES MAINTENANCE SYSTEMS

Another approach would be to have telephone facilities maintenance systems installed at each post as part of the telephone upgrade. This system would maintain the local telephone system subscriber records and plant-in-place drawings. The system would also have the capability of being accessed by a remote computer. In this way, field surveys can be reduced in duration or possibly eliminated. This also would provide the engineers with ready access to traffic data profiles. Traffic data will be used in the next generation of the engineering system to provide more accurate cable and switch sizing.

BIOGRAPHY

Mr. Correll, an employee of Cochise, Incorporated, is currently involved in the analysis and design of computer-aided engineering applications for army communications systems. He is a graduate of the University of Illinois with a B. S. in mathematics and computer science in 1971. He was with the Department of the Army for 13 years as a computer systems analyst working on communications automation before joining Cochise Inc. in 1984.

9

CIMPLEX
BRIDGING THE GAP BETWEEN "COMPUTER AIDED...." AND COMPUTER INTEGRATED MANUFACTURING

Kenneth K. Dellenbach
Automation Technology Products
1671 Dell Avenue
Campbell, California 95008 U.S. A.

ABSTRACT

Until now, all software development has been focused individually on design, analysis or manufacturing, and the information generated by each one of these computer-aided disciplines has existed in independent databanks.

True computer-aided manufacturing requires a new approach to both software design and production. Merely combining the elements of computer-aided engineering, design manufacturing, drafting and documentation, along with information management, does not automatically place a company in the new realm of computer-integrated manufacturing.

CIMPLEX is a product set of interactive software specifically developed to support Computer Integrated Manufacturing (CIM) and to make the transition from "Computer-aided...." to Factory Automation.

INTRODUCTION

Until now, all software developed has been focused individually on design, analysis or manufacturing, and the information generated by each one of these computer-aided disciplines has existed in independent databanks. The industry needed software that could provide engineering tools for computer-aided design, analysis and, at the same time, would permit the engineer to gather and use the data generated by one computer-aided discipline for the next phase of product development.

The productivity breakthroughs in each discipline were producing an abundance of information which could not be easily utilized. There appeared to be no straightforward means of passing and coordinating the information along the critical paths from design conception to production.

Automation Technology Products (ATP) decided to bridge these gaps and develop a product set that makes the transition from what is now termed "Computer-Aided..." to what is considered the next major productivity breakthrough - "Computer-Integrated Manufacturing". But breakthroughs in productivity do not solve the entire problem facing manufacturers. Increase in productivity must be coupled with improvements in product quality and reliability, providing the proper mix of ingredients needed by the manufacturer to maintain a competitive edge.

The product set developed is a uniform complex of modules addressing mechanical parts design, analysis and manufacturing. The three modules are tightly woven within the framework of a single highly advanced database which provides the integration, coordination, control and management of the complete "design concept-to-delivered product" process. The product set is named CIMPLEX. The modules that comprise the set are called CIMPLEX-Design, CIMPLEX-Analysis, CIMPLEX-Manufacturing. These products are complemented by a comprehensive information management module, itself a complete standalone software package, which is called Factory Automation Information Management, CIMPLEX - FAIM.

The product set operates within the IBM VM/CMS environment and is IBM complementary. We believe that being IBM-complementary means much more than being "IBM-compatible." The term "IBM-complementary" means that the software product set brings new and additonal functionality to existing IBM software and IBM compatible software such as CADAM, COPTICS, etc.

CIMPLEX is completely interactive and takes maximum advantage of the latest state-of-the-art hardware advances in graphics display and workstation technology, including all of the newest IBM introductions such as the IBM 5080 series.

The software features major advances in the assistance provided to all of the engineering disciplines, including information management. With non-integrated products, many of the companies with large investments in Computer-Aided areas are currently near overall maximum throughput levels. They are aware that the next significant breakthrough, coupled with improved product quality and reliability, must come from the complete integreation of "Computer-Aided" applications.

A NEW APPROACH TO SOLVING APPLICATIONS PROBLEMS

Up to the present time, the applications software for design, analysis and manufacturing using computer-aided technology was developed separately and independently, either within a single company structure or by independent companies specializing in one of the applications. As a result, there were few if any interrelationships between the applications areas, and each application acquired its own set of methodologies. It seemed that no one was addressing the overall problem of taking the product from design-to-delivery, an area critical to the success of the entire process.

A comprehensive set of supporting application programs that addresses the needs for the automated factory, taking the manufacturing of products from the design stage all the way to the machine shop are now available. All of the modules are, in turn, supported by a single information management program that tracks and organizes the data as it is generated, making it easy to control the production process. Here is an in-depth description of each of these software modules.

CIMPLEX DESIGN

Mechanical Design is the discipline practiced by mechanical design engineers. This module is a tool intended for the use by the engineer as well as illustrators or draftsmen. As a design tool, it is also intended for the manufacturing engineer who must not only make the designed part but who must also design the tools and fixtures required to produce the part.

In the past, "Solids Modelers" have been used as a means of displaying photographic quality imagery of already designed parts. Very rarely is any original design done on one of these systems. The software is simply too cumbersome to use; data input is tedious; and there are virtually no downstream benefits, other than perhaps elegant ways to display parts. The man/machine or, rather, the engineer/machine interface is also cumbersome.

CIMPLEX software has functionality that is matched to the function-ality of the engineer. That is, the engineer is able to invoke his concepts using familiar terms such as Drill, Tap, Define Fillet, Boss, Pocket, Chamfer, etc. This method contrasts sharply to the current use of solids modelers where the engineer has to design in terms of mathematical commands.

But most importantly, designs have a habit of changing, and there are literally no "Solids Modeling" systems that provide any practical means of editing three-dimensional changes. Now this capability is available. For example, consider a simple fillet. Current solid systems use radius information and do an excellent job of depicting a fillet. After that, however, all filleting information is lost, making it virtually impossible to make alterations without having to remove what is already there and beginning again with a new fillet definition.

All this is unnecessary. The software "understands" that the geometry created is indeed a fillet with a specified radius. Thus, if a change of the filleting is desired, the user simply specifies new parameters and the task is automatically completed.

Nevertheless, the ultimate design software tool still falls short unless the information created by the design can be used directly by all of the subsequent engineering disciplines, such as engineering analysis and manufacturing. Without this type of integration, the true downstream benefits may never be achievable.

Providing this integreation requires that the complete part definition be included in the database. The complete part geometry, although necessary is not sufficient. Other non-geometrical data is essential, such as tolerance specifications, radius of curvatures, hole diameters, etc. Without this data, manufacturing engineers can-not possibly know how to machine the parts successfully. Also, the design engineer himself cannot hope to alter his design without virtually starting over again. Since there are limitless ways to make a part, the non-geometric data carried in the CIMPLEX databank along with the complete part geometry not only makes the manufacturing engineer's job more efficient, but enables optimum economics and time savings to be achieved.

The Design software incorporates the very latest advances in mathematics. Surfaces may be defined in any number of ways. A library of over 14 primitive solid shapes, including many complex geometries such as blends, volumes of revolution, swept profiles, and patched solids, can be used to efficiently create designs. The software takes advantage of full Boolean operations, trimmed surfaces, and the like. But all of these techniques are virtually invisible to the user. For example, commands to Drill, or to Place a Fillet, trigger the software to execute the necessary Boolean and other operations that may be necessary to complete the task.

The software is interactive. This means that the time elapsed from issuing commands to three-dimensional visual results is brief. Also, the available wire-frame representations may be continously rotated and scaled in real-time.

Images produced using our software are images of the real part. They are not a likeness or an image generated to create an illusion. What you see is what the object really is, and the information used to display the part is precisely the same information used to analyze and, subsequently, to machine the part. There is only one description of the part. It is a unique description that precisely describes the part, so that the information can be used to manufacturer it.

CIMPLEX ANALYSIS

Analysis is tredionally performed by specialized analysts during a part's development process. The module is intended for use not only by the analysts but also by the designers during the early stages of the product design cycle. In addition to finite element thermal/stress analyses, the software also incorporates interfaces to mechanism analysis capability. These analysis tools form a complete set necessary for most mechanical parts design.

Analysis always required extensive effort on the part of specialists. They had to build analytical models from the designed part for a particular analysis. The usual turnaround time for this is prohibitively long. Consequently, not only was analysis forgotten during the early stages of the product design cycle, it was usually ignored until the part failed during testing or operation.

The major bottleneck of earlier computer-aided systems was the lack of a consistent design database suitable for the analysis discipline. Analytical models, such as a finite element mesh, had to be constructed manually from a sketchy database of a part's geometry, usually consisting of wireframe data. The process of building and verifying the analytical model could take days, if not months, even with the aid of finite element preprocessors. To make matters worse, various analysis disciplines typically used in mechanical parts design, such as finite element analysis and mechanism analysis, were not integrated. Human intervention became mandatory.

Interpretation and conversion of data from design to analysis, or from one analysis to another, had to be done along each step of the way. This complicated the problems and inevitably caused more errors and delays.

The only practical solution is to use one consistent database for the part. It is then possible to automate the generation of an analytical model and to transfer the data between analysis disciplines.

With CIMPLEX software, the engineer designs the part with solids modeling tools. He then specifies the operating environment, such as loading, temperature, etc., directly on the part geometry, not the analytical model of the part. This is far more natural to the designer because analytical models are merely vehicles for analysis programs.

Analytical models should never be the concern of the designer. Because the solids part database contains all the information about the part, the system is able to automatically generate the complete finite element mesh and to convert the operating environment to the engineers.

This facility is available for commercially available analysis programs such as ANSYS and MSC/NASTRAN, and the user may extend this to other in-house analysis code. The designer is completely free to use the system while the system analysis tasks are done in the background.

When the analysis is completed, the software will automatically convert the analysis results back to a form which may be displayed and associate the results with the part for evaluation and for further processing. For example, the temperature distribution in a part resulting from a thermal analysis can be directly used as an operating environment for a subsequent stress analysis.

With the automatic analysis tools provided by the software, there is no need for specialists to perform the analysis. Designers can invoke this function at any appropriate time during the initial phases of the design process. This ensures efficient product design along the design path and allows for any significant product modification at an early stage in the design cycle.

The responsibility of the analysts will be to evaluate the analysis results and suggest design improvements instead of wasting valuable time building analytical models. Productivity and overall throughput in the whole product design process increases. The increase is significant because of the close cooperation between designers and analysts and extends far beyond what is achieved with today's "computer-aided" tools.

When machines that have moving parts are being designed, good mechaniusm analysis is critical to ensure that the product functions properly. The important kinds of information a designer needs from mechanism analysis centers on the motion of the mechanism and the transfer of forces within the mechanism.

CIMPLEX mechanism analysis capability allows both kinematic analysis for the study of the motion of the mechanism and dynamic analysis for force analysis. The mechanism modeling tool uses the solid geometries of the mechanism parts in the design database directly, instead of the traditional stick figures used in today's mechanism analysis programs.

Therefore, not only geometric conversion from a design to a mechanism model is avoided, but the necessary mass properties of the parts can also be generated automatically for dynamic analysis. In addition, the user interface is friendly and natural to the engineers. For example, cylindrical joints are recognized by the system as a boss and a hole mating together in a mechanism assembly.

As in the finite element analysis process, the task schedular handles the execution of the designated mechanism analysis program with input data automatically generated by the system. It then brings the analysis results back for evaluation and further applications.

CIMPLEX analysis is closely linked to DRAM-ADAMS using similar functionality to that described for finite element analysis.

One of the most difficult tasks in the stress analysis of moving parts is to accurately estimate the loads on individual parts in motion. The dynamic analysis of the mechanism is able to predict the dynamic forces acting on ther part much more accurately. These forces can be acceleration of the part and jerking forces acting at the joints.

Since the mechanism analysis uses the samne part geometries as those used by the finite element stress analysis, the mechanism analysis results associated with the part can be used directly as the operating environment for stress analysis. No manual conversion of data is needed between the two analysis disciplines, leading to improvement of product function and design integrity.

Since the analysis results are stored in the database and are associated with the part, engineers can interactively query the results and apply the appropriate design rules to evaluate the part's performance. The information management aspect also provides the long-sought-after tool that can manage the voluminous data generated from the analysis programs. The engineers can retrieve the analysis results of one particular revision of a part under one particular set of operating conditions in order to study the impact of design changes on a part's performance.

The graphical presentation of the analysis results is another major strength of the CIMPLEX software. Color shading of the temperature of stress distribution on the visible surfaces of the deformed part can be displayed.

Users can dynamically orient the part in real time, and the system displays the analysis results interactively at the user-specified view. Animation of the motion of the mechanism assembly or the natural mode shapes of the individual parts is also possible. The user may study the motion while rotating the model in real time. This allows much better visualization of the part's performance, in contrast to the current static displays.

The ability to section the part for study of stress or thermal behavior in the interior of the part is also provided. All these features enable engineers to evaluate the design much more easily and comprehensively; the process is quicker than leafing through reams of computer printouts.

CIMPLEX MANUFACTURING

Making parts that have pockets, fillets, flange-tops and complex sculptured surfaces requires not only complex machines but also talented manufacturing engineers. However, our software relieves the engineers of the contraints imposed by earlier systems. For example, cutter paths may be generated using any free-form cutter shape.

The Manufacturing module is a Numeric Control (NC) processor. It uses cutter libraries, machining parameters, part definition, necessary fixtures, and/or additional tooling features. The result is an NC program and a sequence of cutter positions. The resultant NC program may be further "post processed" to produce unique control instructions necessary to drive a particular multi-axis machine tool. This software module also produces visuals, allowing engineers to verify that the part being made is the part that was designed and, NC programs created from different NC processors, can also be verified. For example, it can produce visuals of an NC program created from an IBM APT-AC processor. The user need only select the necessary interface. The visual sequence produced is unque to the industry.

The verification process begins with the current "work in progress." This is a 3-dimensional visual of raw stock held in place with the specified fixtures. What follows are systematic "scenes" depicting the removal of material.

The material removed corresponds precisely to both the cutter path and the cutter shape. This process involves Boolean subtractions of the tool shape from the "work-in-progress" model. Each new scene becomes the current "work-in-progress" model. Because of color schemes, the engineer can easily detect if material is being removed from the fixture. Similarly, he can detect deviations from the "work-in-progress" model from the original design.

Errors of design deviations and fixture collisions are corrected within the NC program before expensive material, machines and labor have been commited. This is known as "soft" inspection and has proven to be more effective for error detection, part quality and reliability than "hard" inspection, which is both time and material intensive.

Turning, drilling, contouring, profiling, and multi-path (pocket and regional sculpturing) milling capability are provided. This versatility enables interactive 2-to-5 axis NC programming that reduces preparation time and cost. Additional features include:

Free-Form Cutter Libraries

Tool List Libraries (grouping cutters into "kits")

Efficient NC Progream Editing Commands

Selectable Mode Viewing of Cutter Positions

Interfaces to Other Popular NC Processors

Facilities for Process Planning are also provided. Users may create and modify process planning documents for routing and operation. This is accomplished through the use of a predefined library of operation instructions. The documents not only include "instruction" text, but may be augmented with graphical "sketches" of exploded views of assemblies and machine part/tooling set-ups.

Also provided are the facilities to communicate directly to the shop floor through the use of "ruggedized" terminals. Status information is retained in the database and provides for efficient exchange of information, including manufacturing histories, from design to manufacturing.

The manufacturing software module brings to the user the benefit of having one complete part definition. Carrying through only one definition from solids design to multi-axis machining elimenates the barriers and allows the flow of processes that "Computer-Aided. . ." technology alone could not provide.

FACTORY AUTOMATION INFORMATION MANAGEMENT (FAIM)

At the heart of CIMPLEX is an advanced information management system. It is a system that operates on objects and their relationships, whether they be people, parts, processes or even other CAD drawing files.

FAIM recognizes, addresses and provides applications solutions for the toughest of information management problems. This is accomplished through off-the-shelf modules to control engineering change, perform configuration management and centrally control drawing files from various CAD systems. These modules bring additional advantages, such as version control, archiving and intelligent cataloging, to factory automation processes.

For companies with unique needs or existing software that has been developed in-house, FAIM provides the power to add system management capabilities in the most complex situations. Using data modeling techniques that include existing CODASYL conventions, it can be used by a company to coordinate centers of information management.

FAIM software has such features and automated functions as:

Active Monitoring

Change Control

Task Scheduling

Total Interactivity

Total Integration

These features take the conventional relational database management concept to the level necessary to make computer-integrated manufactureing practical. It is an intelligent catalog and provides an integrated and comprehensive information management system for the entire CIMPLEX product set.

Manufacturing organizations need to coordinate activities efficiently. Coordination has direct impact on productivity and profitability. Coordination does exist in varying degrees, but can be found only within isolated pockets here and there. Little if any communication exists between these pockets.

The problem is accentuated by the size, growth and diversity of the manufacturing environment. the trend is to independently automate these dependent activities. This, unfortunately, broadens the information gap. FAIM software provides a solution: it distributes information and knowledge throughout a company in a coordinated and efficient manner. It is a mechanism that enables users to establish programs and routines, to automatically coordinate activities in accordance with predefined rules and procedures.

Rules may be established whereby designs cannot be released to manufacturing without prior approval of the analysis team. Similarly, rules may be defined so that parts cannot be arbitrarily changed without approval from manufacturing engineering. Archiving can be made so efficient that NC tapes, created several months earlier, can be instantly retrieved for a scheduled production run.

But rules by themselves do not solve the problem. People must be notified of changes. There is no other way to trigger procedures and routines. The manufacturing engineer needs to know if "Design" has modified a part before he commits his manufacturing resources. User notification of changes is perhaps the essence of FAIM, for without it, coordination collapses. And sending notification of a change can cue a mandatory response within a given time. If, after that time, acknowledgement is not received, hard-copy notices may be triggered and other alert messages transmitted to invoke other procedures.

Version control, change control, active monitoring and task schedul-ing are the logistics facing all manufacturers. FAIM addresses these issues at the very outset of a design concept and tracks the process through every phase.

The language is unique, friendly and easy to use. It is intended to be used throughout all of the user's activity centers. People from varying degrees of education and background can use it because it is versatile, comprehensive and functional. Almost all instructions are executed simply by pointing to information within a menu. Information within a menu is self-explanatory, especially when the menus are user-defined.

BIOGRAPHY

Kenneth Dellenbach is a manager for Automation Technology Products responsible for customer training and a consultant to R&D software development in mechanical design applications. Previously he was a staff engineer for Amdahl Corporation responsible for the implementation and management of CAD in the Mechanical Engineering area. From 1974 to 1981 at Lockheed Missile and Space Company he advised Advanced System Division personnel in the implementation and use of CADAM and other CAD systems and taught CADAM classes. His 20 years of design experience covers a variety of engineering disciplines and includes several thousand hours of CAD terminal time. He attended Brigham Young University and Northrop Institute of Technology majoring in mechanical and aeronautical engineering.

10

THE IMPACT OF AN INTEGRATED DATA BASE ON AUTOMATION IN MAPPING

William L. Folchi
Synercom Technology, Inc.
10405 Corporate Drive
Sugar Land, Texas 77478 U.S.A.

ABSTRACT

Automation of mapping presents unique requirements because of the way the graphic documents produced relate to the operation of the company. These requirements were difficult to meet in the early stages of evolution of computer graphics. This caused automation to move more slowly in mapping than it did in other pursuits. The introduction of data base management concepts to computer graphics has provided the advance necessary to enable widespread automation of mapping. This has lead to proliferation of distributed graphics systems using more powerful graphic workstations.

INTRODUCTION

Complete automation of mapping procedures like automation of any other activity will only occur when there are sufficient economic incentives. The equipment investment in manual drafting is negligible compared to that for computer graphics. The economic advantage must therefore be derived from substantially improved productivity which offsets the increased equipment costs. This economic advantage has not always been as apparent in mapping as it is in other areas.

BACKGROUND

The application of computer graphics to mapping has lagged automation of other drafting related pursuits. Organizations which have geographically distributed property and equipment; public utilities, municipalities and resource companies deal with tremendous numbers of maps and have an incentive to improve their ability to manage graphic information. A typical public utility serving a large metropolitan area will have between 5,000 and 10,000 hard copy maps to maintain. Revision rates are high because changes in outside plant are frequent and the map is often the only record of the change. It was therefore not surprising that mapping applications were among the first areas in which automation was attempted when computer graphics systems first began to appear.

Mapping however presented some unique problems which made it more resistant to automation than other disciplines. Most significant was the lack of repetition. Computer assisted drafting systems do well when there is some symetry or repetition involved. Detail may actually take longer to create the first time using automation than it would manually. The second and subsequent appearances of detail are almost free to the electrical or mechanical designer using computer graphics. But in mapping all of the effort involves the creation of the detail for the first time and it is rare that repitition of anything except the most elementary symbol occurs.

Maps also have the unfortunate property that annotation can seldom be placed automatically because it frequently requires specialized spacing, size and orientation which change with every instance. Maps also by their nature are graphical presentations of objects whose physical size can not be confined by the limitations of a single drawing. While one would expect this limitation to disappear with computerization most automation approches were designed around the requirements of mechanical and electrical design functions and did not concern themselves with the basic problems of automated map drafting.

Finally in the mapping application the most frequent requirement is to post changes to existing drawings. Except when a new area is assimilated, or a new subdivision is created, most of the activity involves introducing new information to an old document. When the change is small even the high speed plotter has a hard time competing with the human because it usually has to draw the whole map over while the human can add a few lines which describe the change to the existing drawing. In order to have the same advantage the automated system would have to already have the old map stored in a data base. This brings into focus the most serious drawback to automation in mapping, the need to convert huge numbers of existing documents to digital form on the front end. While other graphics applications can frequently begin operating without old information most mapping problems invaribly involve a reference to the existing situation.

These conditions caused computer graphics to be less economical and as a result automation did not occur in mapping as rapidly as it did in other disciplines. The cases where automation did prove advantageous were ones in which there was some fortunate circumstance; the need to produce a new map series at a different scale or a large backlog of low density maps to be revised or possibly the creation of a new base map using automation. These situations still focused entirely on the graphic product and had little impact on the role maps play in the operation of a company.

DATA BASE MANAGEMENT

The most significant development in the automation of mapping was the integration of an attribute data base. This enables the graphical results displayed on a map to be influenced by changes in attributes so that the map will always reflect the current contents of the data base. The resulting computer based system for the management of outside plant records enhances the economic case for automation substantially. Several important operations in facilities management are assisted by the availability of more timely and precise records made possible by an automated records system which contains graphic and attribute data in a single data base. The economic analysis of computer graphics therefore extends beyond efficiency in map drafting and into the operations which use maps. Questions which relate to attribute values but require graphical results are easily answered. For example - produce a special purpose map with the locations of all vacant buildings highlighted, or showing all wooden utility poles installed before 1962 that are greater than 30 feet high. The resulting graphics are easy to obtain by automation but may require so much effort to produce manually that they would not be used.

An automated records system can also produce reports showing installed plant by area. This permits the exact investment in a tax district or a rate area to be quickly available. These operations require so much effort to perform manually that an approximation is usually used or worse the report is not updated when a boundary changes. The savings related to having accurate information showing plant investment by area are substantial and show pay back of the start up cost in a short time - two to three years. Public utility commissions are beginning to require precise data to substantiate rate increase requests.

The availability of a data base to manage graphic data makes it possible to perform other operations beyond record keeping and it is these operations which finally will cause the proliferation of computer graphics in mapping.

The two most common activities which require the use of maps are planning and construction. Both of these activities begin with an understanding of what presently exists and proceed from proposed to under construction to as-built. The following two examples illustrate how computer graphics can be used to enhance productivity by making more information available.

PLANNING

Figure 1a shows the index map of an Air Force base whose master plan has been converted to a digital mapping and attribute data base. The area mapped is 4 by 3 miles. Manual records maintained on 1" to 100' scale and 30" by 24" maps require 28 maps just for the topographic data. Over 65 distinct map products are maintained. In the computer sense the area is mapped and stored as a continuous digital map in ground units (feet or meters). The scale, content, and area of a map product is determined when the product is required. Figure 1b shows the planimetric detail of a small area of the index map. The storm sewer model is shown for a still smaller area in Figure 1c.

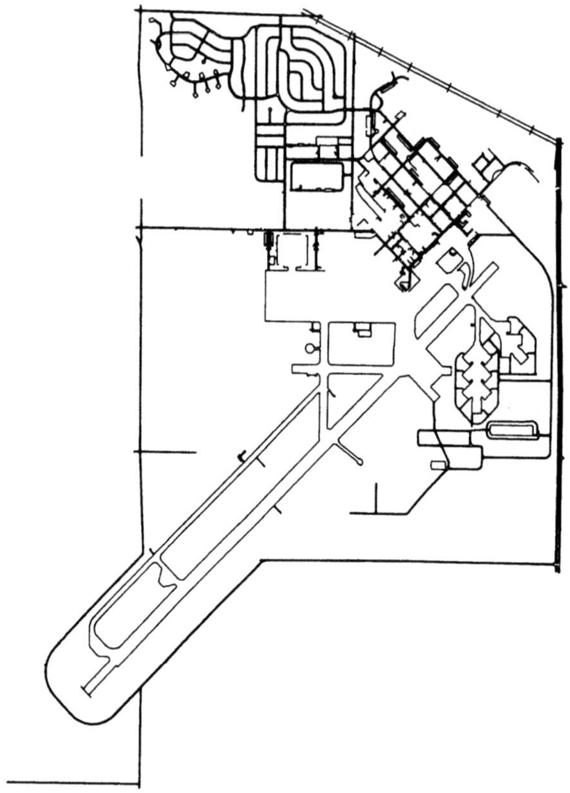

Figure 1a
An Air Force Base Comprehensive Planning Map

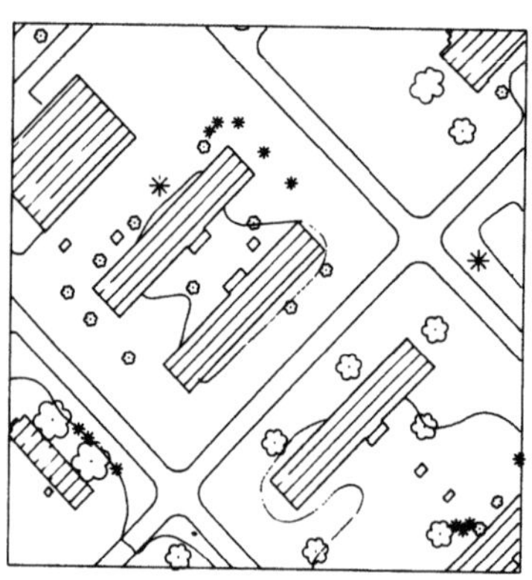

Figure 1b
View of the Data Base
Showing Planimetric Detail

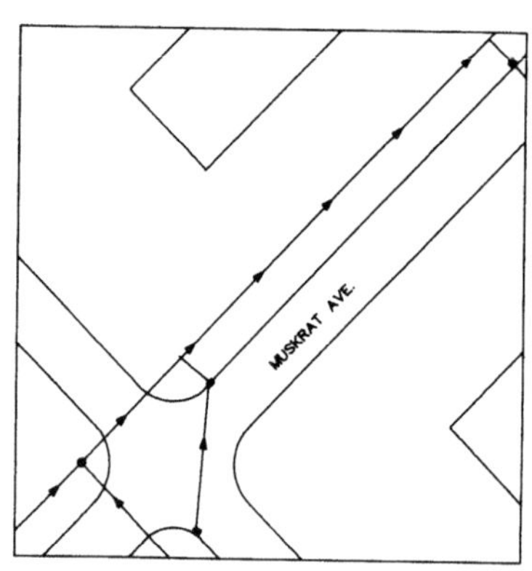

Figure 1c
View of the Data Base
Showing the Sewer System

The data base is organized into the following 27 data types:

Street Centerlines
Contours
Poles
Buildings
Spot Elevation
Transformers
Sidewalks
Runway System
Switches
Trees
Land Use Polygons
Electrical Condutors
Street Names
Proposed Land Use Polygons
Cathotic Protection
Hangers
Capital Improvement Plan
Water Valves
Ammo Dumps
Water Mains
Heating Symbols
Heating Lines
Storm Sewer Symbols
Storm Sewer Mains
Sanitary Sewer Symbols
Sanitory Sewer Mains
Liquid Fuels

Each data type has associated with it a series of attribute records which describe individual elements. For example the building data type include the following attributes:

Facility Number
Area
Land Use Code
Area by Land Use Code
Condition
Status
Disposal Action
Vacant or Occupied

The master planner combines this rich data base with system analysis tools to reduce the effort required and to increase the accuracy of the final plan. For example the planner is required to place a new mess hall. First the attributes of the building are examined automatically. The buildings which support the mess hall, the buildings which house the users of the proposed mess hall and other buildings not affected are crosshatched with different patterns (Figure 2). This helps the planner visualize the situation. The planner may then examine the land use zones as well as other

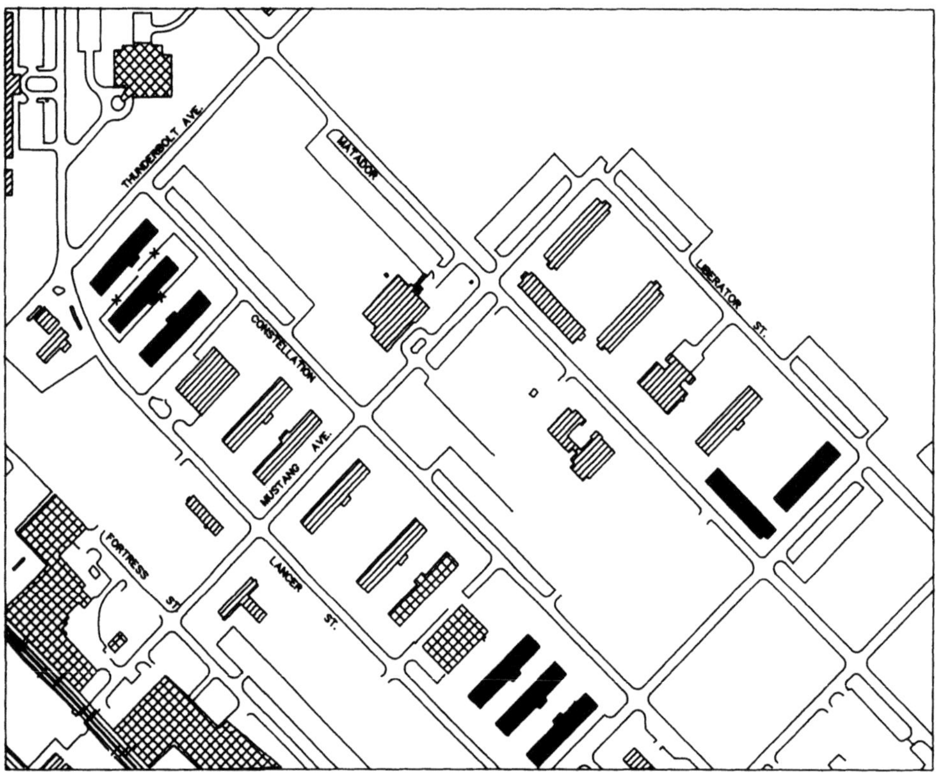

Figure 2

A View of the Planning Data Base With
Buildings Crosshatched According to Their Use

restrictions to choose a tentative location. Finally the planner may display the facilities to estimate the cost of the support for the new building. As the plans are completed the required reports are produced automatically. Figure 3 shows an automated SAC 246 form.

```
                       S.A.C.   246
SUPPORTING DATA PROGRAMMED INSTALLATION     DATE      REPORTS CONTROL
  END MISSION REQUIREMENTS JACKSON AFB, CTG 19-Oct-84 HAF-PRE(AR)  7115
SECTION A
1 CATEGORY  2 NOMENCLATURE        3 AGENCY   4 UNITS     5 QNTY. REQ.
  721312       DORM AM PP/PCS-S       SAC        SF           146665
6 EXT. UNUSABLE    7 EXISTING USABLE 8 NOT IN INVENT 9 CURRENT DEF.
    118001  SF          152664   SF        0                 5999   SF
10 PROGRAMMED AMT.      11 PROG. DEFICIENCY    12 PROG DEFICIENCY COST
       0                      0                       0
SECTION B                 COMPUTATIONS AND SOURCE DOCUMENTS
1 COMPUTATION OF REQUIREMENTS               2 AUTHORIZATION OR SOUR.
  COMPLIMENT 18 MEN EQUALS 5700 SF             AFM 86-2, PARA 16-10C

  MOBILITY EQUIPEMENT STORAGE 3150 SF




SECTION C
EXISTING FACILITIES PRESENTLY USABLE OR PROGRAMMED TO BECOME USABLE
                                                        DISPOSAL ACTION
FACILITY             SCOPE                   STATUS      PROCESS
NUMBER                                       VAC  OCC      YES     NO
0301        17159  SF                         X             X
0302        17159  SF                         X                    X
0304        16116  SF                              X               X
0305        17159  SF                              X               X
0307        17159  SF                              X        X
0309        16090  SF                         X                    X
0311        17159  SF                              X               X
           118001  SF

2           EXISTING USABLE FACILITIES
FACILITY             SCOPE             STATUS        EXISTING  PROGRAMMED
NUMBER                                 VAC  OCC      CODE          CODE
0327        26035  SF                   X            0
0328        25185  SF                        X       2
0329        26035  SF                        X       2
0331        25039  SF                        X       1
0332        25185  SF                        X       1
0333        25185  SF                        X       1
           152664  SF
SECTION D   FACILITIES PROGRAMMED FOR ANOTHER USE
FACILITY             SCOPE            FUTURE NOMENCLATURE  FUTURE  FUTURE
NUMBER                                                     CATEG.  CODE


REMARKS                                               COORDINATOR
                                                    UNIT COMMANDER

                                                    ORGANIZATION
                                                    305 CSG/SP
                                                    CATEGORY CODE
                                                      721312
SAC        246                   SYNERCOM INFORMAP II SYSTEM
```

Figure 3

A Computer Produced Facilities
Status From (USAF-SAC Form 246)

Planning of this type has involved primarily analysis of existing and proposed existing elements (i.e.; building, electrical conductors, etc.). A digital maser plan may also support analysis of spatially organized data. For example the area may be classified in the following coverages:

Land Use
Suitability Corridors about Roads
Environmental Restrictions (Endangered Species)
Historical Sites
Errosion Factors
Equipment Limitations

The planning problem is to determine the suitability for various types of maneuvers. The planner may union these coverages to produce an integrated coverage. Each polygon in the integrated coverage is classified by the attributes of each individual coverage. The classified data is then symbolized to produce the suitability model in Figure 4.

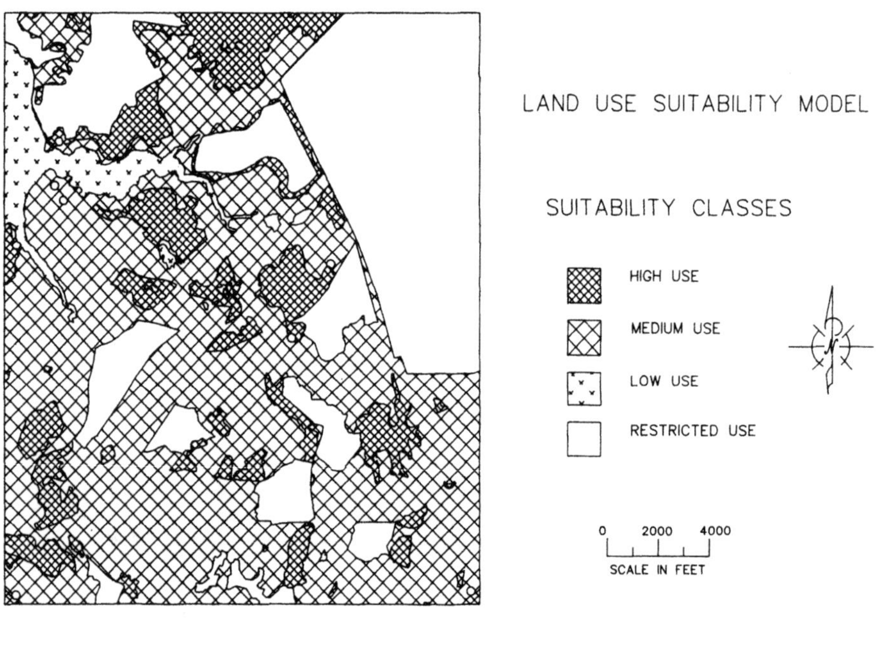

Figure 4

A Land Use Suitability Model Produced From the Union of Classifications Which Effect Suitability

FACILITIES CONSTRUCTION

In this case a telephone service must be extended to serve a new residence. The engineer using the system may first inquire to determine if the nearby cable has additional vacant pairs that are not assigned. He may then place the new cable and symbolize it to show that it is proposed. The system obtains the pair number and pair count range from the attribute record associated with the existing cable and propagates any changes or cable throws along the network. As the design is created a bill of materials showing equipment added and retired can be produced and as well as an estimate of the cost of the job. Also, transmission characteristics of the cable pairs can be produced in a tabular report, commonly known to the telephone industry as a "loop make-up". This report will produce loss and resistance parameters of the pairs along with the length, bridge tap, and terminals affected. In subsequent activity the job may be tracked from its proposed state to under construction by the appropriate data base transactions. All proposed work is always done in working storage (scratch-pad area) so as not to affect the permanent data base until posting of the job order is performed. When the project is completed it can be automatically posted to the permanent data base. This posting function will automatically change proposed symbology to existing, and will reflect all non-graphic information changes associated with the affected items of plant. Figure 5 shows a portion of a work print which which is the document used in the telephone industry to manage outside plant construction.

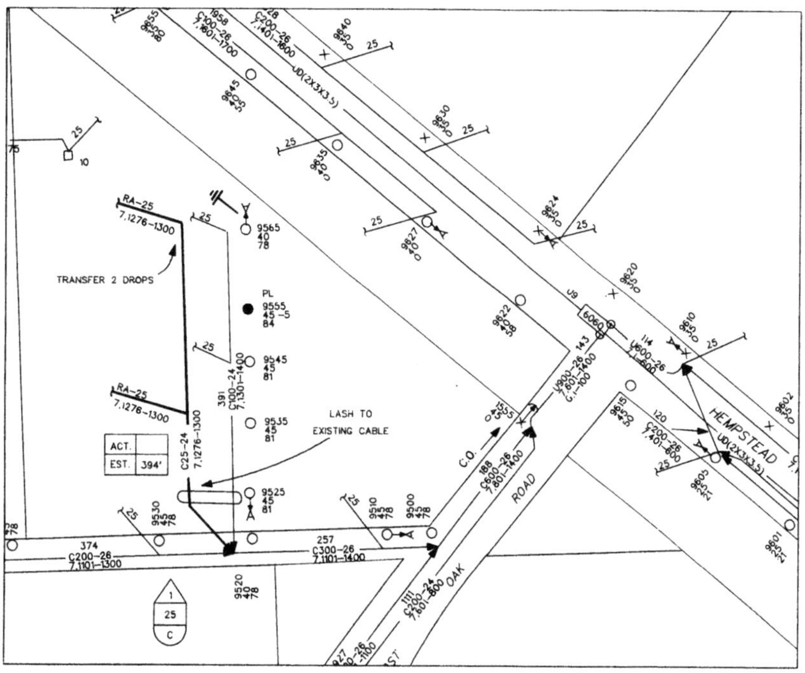

Figure 5

A Computer Produced Outside Plant Engineering Work Print
Pairs 1276 - 1300 of Cable 7 Will be Used to Serve the New Area

In a network application such as the one described a small change may effect the electrical properties of cable several miles away. In a manual sense this might mean the need to revise several maps. The computers ability to deal with the network as an entity and to model the changes throughout enables current information to be available instantaneously from one end of the system to the other. This reduces the effort needed to complete the current design and reduces the research required for other engineers working on different parts of the system. This level of enhancement make the economic case overwhelming in favor of automation. It offers a substantial savings in drafting effort and a larger savings in a much more costly commodity - engineering effort. This reduction in engineering effort which in one telephone company amounts to a reduction of $4.2 million annually is the cause for the recent high interest in automation by telephone companies. Now for the first time an engineer can sit at a workstation and without moving around to look for information or waiting for results from the drafting room he can complete a design and pre-post it automatically to the company data base. This saving collapses weeks into hours. The automated approach offers other benefits and savings which include:

- Improved record accuracy.
- Reduced 'open' job order time (Telephone Plant Under Construction).
- Better retirement records.
- Improved depreciation studies.
- Manpower reductions.
- Ability to interface to other mechanized systems.
- More timely access to existing data.
- Provide graphical multi-year budget summaries.
- Easy tie into central office and building records.
- Improved 'posting' time at job order closing.
- Ease in correcting 'copy' errors and reconciling 'duplicated' record information.
- Improved feeder/distribution design.
- Step towards 'paperless office'.

DISTRIBUTED GRAPHICS

There is substantial incentive then to create a data base and place a workstation at the disposal of all engineers involved in outside plant engineering. However the graphics stations used in the mapping industry are expensive and when the other economic factors such as conversion are rolled into the cost the picture begins to fade again because of the number of stations required to support engineering.

Fortunately technology has once again made a quantum leap and at present we see low cost engineering stations with the required computational power just over the horizon. Installations are just beginning to be made in mapping applications and it may be a little early to be making firm predictions but the picture certainly is favorable at this time. The first point, that outside plant engineers can derive substantially increased productivity from the availability of an intelligent graphic data base, has been proven in one case in point Continental Telephone. Continental began with a fast track approach by going directly to their distributed engineering offices. The central processor is located in Wentzville, Mo with stations in Wentzville, Bakersfield, Ca, Victorville, Ca, Gray, Ga, Crockett. Tx, Dallas, Tx, Atlanta, Ga, Pell City, Ala. A pilot phase was conducted using the current technology equipment and software. Continental found that with all of the problems expected in taking on project of this magnitude they were still able to create a representative data base for each of three operating divisions; assimilate the technology into their operation and derive an economic advantage in their construction activities within one year of installation. In future progressions additional CPU's will be added to separate the regions and lower cost stations will be added to support engineers and other users.

CONCLUSION

Integration of a data base and powerful analytical functions with computer graphics has enabled the automation of mapping related operations on a cost effective basis. This has created an environment in which engineers interact directly with the information previously stored on maps and perform their work more efficiently. Looking to the future one can see an evolution in which more power and functionality are moved out to the station and the user. Mapping applications however, are so integrated with the operation of the company that they will always require the link to corporate data base.

BIOGRAPHY

Bill Folchi is Vice President of Synercom's Federal Systems Operation. He is responsible for sales of Synercom's Mapping Information Management Systems to the Federal Government in military and civilian applications. He has been with Synercom for seven years.

In the preceding fifteen years he has pursued a career in computer graphics, mapping and photogrammetry. With Keuffel & Esser Company he designed and developed computing systems in support of computer graphics and mapping products. With Utility Data Corporation he developed and implemented geographically-oriented information management systems which were used in the gas transmission industry, public utilities and land records during the pioneering era of the AM/FM industry.

11

UPDATE OF MASS STORAGE SYSTEMS FROM THE USER'S POINT OF VIEW

J.K. Gingrich
ALPHAREL
Camarillo, CA.

In 1985, procurements by the Army and Air Force of automated engineering storage and retrieval systems are underway: DSREDS, the Army system (Digital Storage and Retrieval Engineering Data System), and EDCARS, the Air Force system (Engineering Data Computer Assisted Retrieval System). AT&T is the contractor; Alpharel is the major subcontractor. DSREDS and EDCARS establish complete drawing, storage and retrieval systems for these sevices and will facilitate the distribution of bid packages. The first installation will be at MICOM at the Red Stone Arsenal in Huntsville, Alabama, and the second at SM/ALC at McClellan Air Force Base in Sacramento, California. Ten other installations will follow. The program includes twelve sites, each requiring approximately 25 terminals, the capacity to store 400 gigabytes, and 400 gigabyte backup, with aperture card scanners, page scanners, large scale scanners, and computer output microfilm (COM) and hardcopy capabilities. Because the DSREDS/EDCARS system is so significant and embodies all attributes of an outstanding automated drawing system, a brief description of it will be given here.

The Army and Air Force maintain massive drawing data bases at these sites, primarily in the form of aperture cards. Drawings included in bid packages for procurements are sent out in aperture card format, resulting in substantial dollar savings in paper and mailing costs. However, the procedures to retrieve and duplicate these aperture cards are labor intensive and time consuming, and therefore a more cost effective and efficient system is necessary. The DSREDS/EDCARS system will significantly reduce the labor and time required to issue a bid package to industry.

The primary input device is the laser aperture card scanner. Two scanners will be located at each site. One will be indexed for an "A" to "C" size scan (16X) at 200 lpi, and the other indexed for "D" and "E" size scan (30X) at 200 lpi, although either scanner may be indexed to scan either size. The scanners have an automatic card feed with a capacity of 500 cards for the input hopper.

During the scan process the card is first fed into a holerith reader stage where the punched data is read. This establishes the retrieval address for the image data base resulting from the aperture scan. It also provides the information for the data base file. If there is a difficulty with the holerith punching, such as a missing required field, the card is rejected to a reject file and no scan takes place. This reject hopper is also used to reject a "D" or "E" size card if it is placed in the "A" to "C" size scanner, and the opposite is true for the other scanner.

Next, the card moves to the scan stage where the "A" to "C" size aperture cards are scanned in approximately 2.5 seconds, "D" and "E" in approximately 10 seconds. Each scanner includes an integrity control circuit station for automatic threshold adjustments for image enhancements. These circuit stations also provide verification of film content by white and black pixel count thresholds to test the digitized data to eliminate blank cards, dark or clear. Such cards are rejected to a second reject hopper (which can also be set to reject cards based on a specific field or fields of the holerith punchings, at the operator's option). Scanned image data moves to a compressor module and is stored on a magnetic disk buffer, awaiting a Q.C. check or further action. The DSREDS/EDCARS system also includes a page size scanner which has an automatic paper feed capable of handling 50 sheets. Each page is scanned at 200 lpi in approximately 3 seconds.

The image data is ultimately stored on a 200 disk capacity mass storage unit with two drives. Each unit is capable of storing up to 400 gigabytes. Disk transfer and spin-up takes about 8 seconds for data retrieval. Two units are located at each site, one to store the data and the other to provide auxiliary optical read/write capability and to serve as a full backup and media copying device.

Data is displayed on a 19-inch diagonal display terminal with 1700 x 2200 per inch resolution. The display can be rotated for landscape or portrait orientation. Pan and zoom capabilities allow a user to quickly and easily view any part of a drawing including an "E" size image.

The display terminal also includes an excellent revisory capability which rivals any 2-D vector system. Moving, erasing, or copying the pixel data is accomplished by manipulating a mouse, an integral part of the terminal. A window menu allows the user to quickly choose the desired activity. Other capabilities include adding lines, curves, splines, and free-hand lines, windowing, and split screen. This revisory capability will be used by the Army and Air Force to revise in-house drawings resulting in drafting time savings.

Bid package aperture cards are obtained by a laser COM, a companion to the laser scanner. As the image data is retrieved from the optical disk, it is sent to the laser COM and the image is imprinted on the film. The COM has high resolution capability and produces "A" to "E" size images at 200 lpi as scanned. Three COM's are required to meet an output requirement of 3000 images per hour.

The resulting roll film is processed and placed on a roll-to-card duplicator. Because the roll-to-card duplicator is on-line to the central processor, it receives holerith data and other instructions such as number of cards (sets) required for each bid package. Applicable data is punched on each card during this duplication procedure.

Bid sets are then reproduced to obtain the desired number of sets through card-to-card duplicators providing an output of over 9000 duplicate cards per hour.

This description is a brief overview of the DSREDS/EDCARS systems and does not include the many details of customization to Army and Air Force procedures. It is intended to provide a basic explanation of how the optical disk storage technology results in a savings of time and money and allows these users to efficiently accomplish their important tasks.

This essentially is a complete automatic system (see Exhibit 1). The only manual intervention required from the time the bid package is generated to the actual bid sets is in two areas. First, the resulting COM film must be removed and placed in a film processor and, after processing, into the roll-to-card duplicator. Second, the disks from the roll-to-card duplicator must be merged into a single discreet bid package. (A third manual effort is required if collators are not included as part of the card-to-card duplicators.)

This DSREDS/EDCARS program is also very important to all potential users. Until now, industry advocates have spoken bravely about the promising new technology that was available, but there was no evidence that potential users were listening or would acquire and employ large advanced systems. With installation of the AT&T systems, this technology receives practical application and many users now see that the technology is practicable and ready to advance.

General Electric Turbine Operations in Schenectady, New York provides another example of an installation of such a system, and GE may lead the way to further advances. They are currently installing a complete system and card scanner. The system will be operational by the beginning of 1985 although it will be the end of the year before the entire data base is fully loaded.

The GE system includes 16 display terminals located in remote locations within one building. Five of these terminals will have raster change capability. Drafters will be able to use these terminals to modify the scanned drawings and create new drawings if necessary. GE also has six electrostatic plotters dispersed throughout the area for hardcopy output and expects to reduce its normal hardcopy printouts by 50%, by having the display terminals available at strategic locations throughout the plant so that engineers, drafters, manufacturing people and others can view a drawing without resorting to hardcopy. GE expects to place at least four terminals in the manufacturing shop environments. As their system grows during the next several years, additional terminals will probably be added with a build-up to reach 35 or more.

GE Turbine Operations will also install a similar system at their plant in Lynn, Massachusetts, and the company plans to transmit drawings digitally from one plant to the other. The GE installation appears to be the first bonafide optical disk-based engineering drawing storage and retrieval system in the U.S. and most likely the world.

There are less advanced automated systems in operation in industry today. An example is another GE system in Salem, Virginia which uses 16mm roll film. There, all drawings have been filmed on 16mm film and are retrieved from roll film. General Electric drive systems can accomplish this because 99% of their drawings are "C" size or smaller. The films are stored in cartridges. As a drawing is requested, the cartridge is retrieved automatically and the film is indexed to the requested aperture on the roll. This aperture is then digitized and transmitted to remotely located terminals. Once the drawing is digitized, it is stored on a magnetic disk for quick retrieval and is available if more than one user wants to view it or obtain a print-out. At a certain point, the drawing is erased from the magnetic disk.

Users with a number of "D" and "E" and larger drawings could not use this 16mm film procedure because adequate resolution would not be attained. The supplier claims that it will build a 35mm system if asked to do so. Pratt and Whitney is said to be considering such a system. While this is a less expensive system than the optical disk-based system, it has several disadvantages. Retrieval on the roll film is done sequentially, and continual use of film could be a problem. Most users have an apertures card data base, so that the changeover to roll film would be time consuming and costly, more so than scanning a card to a digital format.

There are also automated storage and retrieval systems utilizing the aperture card as the base, of which Infodetics and Access are the two major suppliers. Each of these companies claims that its aperture card-based system can be readily adapted to the optical disk-based system, and such is the case. A combined aperture card/optical disk sytem is not unsatisfactory but is expensive.

A number of companies now are marketing a complete optical disk-based automated drawing and storage system. Alpharel, Impres, Integrated Automation, and TERA are among the leaders, with others ready to join, such as Infodetics and Access, and later, UBBM and Kodak.

The aperture card scanner provides a resolution of 200 to 300 lpi (even for "E" size drawings at 30X). Most of these scanners are the charge-coupled device (CCD). The Alpharel model features a laser scanner which provides a superior resolution, and has the added feature of providing a compatible laser COM for high resolution output or raster data to roll film for aperture card generation directly from the raster data. This laser COM is an important part of the DSREDS/EDCARS contract because of the Army and Air Force requirements for large volumes of aperture cards sent out as bid packages.

Hard copy scanners from "A" page size up to "E" size are also available. Such companies as Docuscan and Dest have good page scanners, and Industrial Vision Systems (IVS) and Skantek have machines which can scan up to "E" size at 200 lpi.

There has been concern about the resolution of a drawing when it is viewed on a display terminal, a concern about limitations of the display terminal itself. For example, the 1700 x 2200 pixel resolution display terminals used by the above suppliers limit "E" size drawings to a view capability of 50 lpi. However, through averaging algorithms, the resolution problem is being solved and many "E" size drawings can now be easily recognized. While small details such as dimensions cannot always be seen, they can easily be read by zooming into the drawing further.

Therefore, the technology exists to provide a digital representation of a drawing with good resolution, to display it on a terminal, and to convert it to a film aperture card or a hard copy of the drawing on an electrostatic or laser plotter. A laser plotter will be available to produce up to "C" size. The laser plotter has distinct advantages over an electrostatic plotter because it is faster and prints on regular bond rather than the relatively expensive coated paper which the electrostatic plotter requires. Suppliers of electrostatic plotters advertise speeds up to 2 inches per second, although outputs to meet the speed are not often seen, especially for fairly dense "E" size drawings. But one can expect at least 2 inches per second some day, a rate which will provide a "C" size drawing in 8.5 or 11 seconds depending on the format. The laser printer will print a "C" size in about 5 seconds.

The storage media are now fairly well defined, but there remains the need for a standard--14 inch disks are being produced. Most of the current ones are sized at 12 inches--such as 3M, Thomson CSF, and Phillips disks. Japanese disks are mostly 12 inches, but there are also smaller disks, and the sizes for these have not been standardized. It appears that Europe and Japan are very interested in standards for the entire disk system and are actively working on this problem. U.S. producers do not seem to show the same sense of urgency. There are several good disk and drive suppliers, Thomson CSF and Phillips, in particular.

And the juke box is now a reality. Filenet has a 64 disk cabinet with as many as 4 drives and is building a 200 disk juke box with up to 4 drives for the DSREDS/EDCARS contract described above. Integrated Automation (IA) has a 50-100 disk juke box with an option for 2 drives at the 100 disk level. An IA juke box is in use at the Library of Congress in Washington, D.C., and the redesigned box will be installed at the GE Schenectady site shortly. It is said that CYGNET has a 100 disk juke box also.

Therefore, a complete system is now available and is a viable solution to meet the needs of the user in industry and government. Users are interested in procuring such a system at the least possible cost. Consider the cost of storing a document in digital format as compared to storing that document in any other format. There is no storage medium as cost effective as microfiche at 98X and above. As mentioned earlier, most engineering drawings cannot be reduced to that level. Using 35mm as the base, the costs are greater than those of digitally stored drawings. While the optical disk costs are higher than the 35mm costs, they will soon be lower. Prices for optical disks are $200 each and will fall to $80 and possibly to $50 within a few years. The magnetic optical disk expected on the market in the future could surpass the optical disk as the lowest cost medium, but this will not occur before 1990 or so. Its major characteristic is that it is erasable, but that is not necessarily an advantage. The data on the following chart are for media only and do not include the required associated hardware costs.

When considering an entire system, cost estimates are difficult to make. They vary with the number of display terminals, printer, and other peripherals required. For comparison's sake, one may take an optical disk system of 50 disks in a mass storage unit only, forgetting the peripheral requirements. In this case:

1 mass storage unit (juke box):	$100,000
50 disks at $200 each:	10,000
	==========
Cost to store 500,000 drawings:	$110,000

Assuming 10,000 average drawings per disk, the cost per drawing stored is .22 cents.

On average, the cost of an aperture card is 50 cents for a silver halide drawing, and 5 cents for a diazo copy. This does not include costs of storage cabinets.

If a 2-sided disk were used, which is now becoming predominant, costs would be:

1 mass strage unit (juke box):	$100,000
50 disks at $400 each:	20,000
	==========
Cost to store 1,000,000 drawings:	$120,000

Cost for each drawing: .12 cents.

To carry this one step further, one may assume that a complete system capable of storing 1,000,000 cards would cost 1.5 million dollars. The system would include approximately 16 display terminals and 5 printers. With the double sided disks,the cost is approximately $1.50 for each drawing stored, and this includes the automatic storage and retrieval, with additional substantial labor and material savings.

To keep costs as low as possible, users are asking for an off-the-shelf system. This is a challenge to suppliers, since each user has specific requirements that must be considered and provided for in customized ways. Evaluations of user requirements is itself a challenge. In one case, it was reported that there were 500,000 "A" size cards and this was calculated into scan time and storage space. In a status review, it was discovered that "A" size cards actually included 3 "A" size drawings. Storage requirements for the 500,000 cards quadrupled because of this, and scan time doubled since all of these cards had to be scanned at "C" size rather than "A" size. In another case, a company requested certain data output but did not clearly define that they needed a card-to-card duplicator to be on line with the main system. This required additional software development. Both of these cases were discovered after the original contract had been signed, and were the result of not clearly stating system requirements in advance. The user needs to incorporate a certain amount of program management in the contract, and for a moderate sized system, $50,000 is adequate. This would provide for a meeting of several days at the user's site to ensure that operational procedures, data base requirements, and other factors were well understood. As has been said, users appear to be ready. One other factor must be considered, which is that with increased use of CAD or IAG generated drawings or CAD generated design packages, the need for a drawing tends to be eliminated. Users are claiming that as much as 75% of all new work is now being done on the IAG terminal, and some are claiming that nearly 100% will be done on IAG as early as 1986. They are also becoming more "vector oriented", and their interests in the raster to vector conversions are therefore becoming greater. Several potential users have said they would prefer to store everything in vector rather than raster format. Vectors seem to be better behaved, and the IAG systems are much more sophisticated for purposes of data generation and data revisions.

The raster to vector conversion system is far from perfect, but some good results can be obtained even with today's technology. It is reasonable to assume that most of the difficult problems will be solved in the near future so that raster to vector conversion will be viable. One of the problems to be overcome is the relative inaccuracy of the manually drawn line versus the reasonably absolute accuracy of the IAG drawn line. For example, a drafter may draw a line 2.55 inches, and depending upon how he or she feels at that time, that line could vary from 2.50 to 2.60 inches. If the line is vectorized, it is vectorized at 2.5 or 2.6 inches, instead of 2.55 inches, and placed in an accurate data base inaccurately. One company has claimed that they can correct this problem, but that remains to be seen. At any rate, the heavy use of vector images will impact the users. Some who are now considering large investments in a raster system might instead remain with vector systems because a large percentage of their active drawings will be in vector format.

The subject of retrieval then becomes pertinent in considering the problem of storing raster, vector, and IGES data. In the recent past, a basic problem was the inability to quickly display or obtain a hard copy of vector image. The vector data had to pass through a slow vector-to-raster converter before they could be printed on an electrostatic or laser plotter, or displayed on a display terminal other than the originating IAG terminal. It was recommended that two files of IAG generated data were required, one for the vector data for the IAG users, and one raster file for rapid display at remote non-IAG stations and for output to electrostatic plotters.

In 1985, this is no longer the case. The Alpharel display terminal, for one, can display both vector and raster data, and can display these data interchangeably. The vector to raster conversion works quickly enough that an electrostatic or laser plot is viable directly from the vector file. Therefore only one data base is required for IAG generated drawings.

This system provides an excellent advantage to the Army and Air Force. Several sites will be receiving IAG generated drawings in magnetic tape format instead of aperture cards. This practice will probably increase substantially over the next few years as more and more drawings are generated on IAG. Such tapes can be entered directly into the optical disk file as vector data and be available for retrieval at a display terminal, on the COM for aperture card output, or at a hardcopy unit, a laser, or electrostatic plotter. The only requirement is that the IAG output must be in one of the standard output formats such as Cal Comp out or Gerber out.

Therefore, the need for the optical disk storage and retrieval is still significant even if a user tends to go "all vector." The only feature that might be affected is the raster change capability. Some users may decide to run all changes using vector rather than a raster change.

The future for the optical disk engineering drawing system looks bright. DSREDS and EDCARS installations are progressing and, by the end of 1985, testing will begin at the first site. Interest in these installations and the GE Schenectady program is very high.

BIOGRAPHY

Jack K. Gingrich

Mr. Gingrich, after graduating from Lafayette College in Easton, Pennsylvania, joined the General Electric Company where he spent most of his 35 years in various engineering department services work. In 1968, he accepted the position of Manager, Engineering Services and Administration in the Medium Steam Turbine Department. Among other duties, he was responsible for all of the functions associated with drawing and technical manual reproduction. He initiated, procured, and managed the first interactive graphics system for the Medium Steam Turbine Department. He later participated on the GE Corporate study team on Data Base Management Systems and led the study team in automated drawing storage and retrieval systems. As a result of these studies, one General Electric Department has placed an order for an optical disk storage system with many others to follow. In June 1983, Mr. Gingrich retired from the General Electric Company to go into consulting work and has recently agreed to act as a consultant to ALPHAREL, Incorporated.

12

PLANNING PRECEPTS FOR CAD INSTRUCTIONAL DESIGN

Cecil L. Groves
Houston Instrument
P. O. Box 14547
Austin, Texas 78761 U.S.A.

Abstract

Employer demands for persons trained in the use of computer-aided drafting (CAD) has exceeded most education and training projections. In trying to meet these demands, educational institutions have encountered significant instructional obstacles. As an example, publishers who provide instructional materials in support of the different teaching disciplines have deferred publication of CAD instructional materials except at the most general level, because of the hardware-specific orientation of CAD. Colleges and schools have been forced to rely on CAD vendors to provide the necessary teacher training and instructional materials. This might be all right except that CAD vendors have focused their research and development, as well as application support efforts, at the commercial market and not the education market. The primary purpose of this paper is twofold. First, issues considered critical to CAD instruction at the secondary and postsecondary level institutions, who are responsible for training operators and technicians, are discussed. Second, given the preceding critical issues, instructional design principles and observations related to CAD instruction at the secondary and postsecondary levels are presented.

1. INTRODUCTION

Present and future demands for trained computer-aided drafting/design (CAD) operators exceeds the capabilities of our educational system to meet those needs. This is an inherent problem of American education resulting from our lack of a specific manpower planning policy, which places our educational institutions in a reactionary posture of responding to already existing manpower needs. Other occupational areas, where computers and telecommunications have revolutionized traditional occupations, face the same problem.

The advent of CAD has impacted technical education in the last three years more quickly and completely than have major technical advances in other occupational areas. For example, in machining and office occupations, where automated systems such as numerical control and word processing have been introduced, the major changes began in the late 1960s and have steadily escalated. When computer-aided drafting/design dropped in price under $100,000 in the early 1980s, there was an explosion of activity nationally as small and mid-sized industries and businesses began to convert one of the oldest and most stable occupational areas--drafting--from a labor intensive activity to an information-based, computer-managed occupation. The drastic change of such a fundamental vocational skill was further shocked by the shortened timeframe imposed for providing the workplace with a workforce trained to use this new drafting tool.

Colleges and schools were suddenly expected to spend tens of thousands of dollars in acquiring automated drafting systems. This was more money for equipment than most schools and colleges had spent cumulatively in the past 20 years. In addition, traditional drafting faculty were asked to become knowledgeable concerning information-based, computer-managed drafting and responsible for the purchase of the necessary equipment. Further, as has too often been the case, responsible faculty were asked to do this in their spare time as an additional assignment. If this was not difficult enough, once the CAD equipment arrived, approved course outlines and syllabi were expected with established laboratory assignments, plus, someone eager to teach the course. Little wonder educational institutions and their drafting faculty are experiencing difficulty in responsibly meeting CAD manpower training demands.

2. INSTRUCTIONAL ISSUES AND CONCERNS

Educational institutions, too often criticized for not being responsive or relevant, are making a determined effort to introduce CAD into their drafting curricula. While such responsiveness is certainly commendable, it has led, too often, to the purchase of CAD equipment inappropriate to the instructional needs of the institution. It can be argued, with some justification, that the seeding of a new technology in vocational-technical education is acknowledgeably expensive and often characterized by trial and error. However, the colleges and schools are painfully coming to realize that the purchase of CAD equipment should be based on the level of skills--CAD familiarization, CAD training, or CAD design--being taught and not simply the availability of CAD equipment at a good price. This new realization is being caused by more and more states, due to the large expenditures for CAD equipment, applying accountability criteria such as cost-per-unit-of-instruction, student job placements and employer satisfaction, and instructional practices. For the past year, Bausch & Lomb sought input from numerous drafting educators and occupational instructional specialists in identifying critical issues and concerns for CAD education. Some of these issues and concerns are discussed in this section.

ACCOUNTABILITY

Urging along colleges and schools to address the need for training CAD operators/technicians has been the availability of funds from state, local, and federal sources. States have been strong supporters of CAD training programs assuming the availability of such training in their public educational institutions would be an attraction for industrial relocation and development. Increasingly, states are beginning to demand an accounting for the funds spent on CAD. Usually, this accounting is in terms of number of students taught, job placements, and overall cost-per-unit-of-instruction. As accountability criteria are applied to CAD instruction, institutions have begun to seriously consider the instructional issues associated with CAD. Unfortunately, this consideration often comes after the purchase of the equipment.

EQUIPMENT SELECTION AND INSTRUCTIONAL APPLICATION

When CAD was first introduced into two-year colleges, many institutions purchased systems which required operators to have programming skills and structured their curriculum around this requirement. Today, programming skills are not necessary, or required, for basic CAD operation. Other institutions acquired powerful design-oriented CAD systems (usually at a significant price discount) only to use a small portion of the equipment's capabilities for instructional purposes since students lacked the math and science backgrounds for the engineering-oriented operation. In a final example, most institutions have purchased micro-based CAD systems which emulate the industry-focused CAD systems. These micro-based systems were low-cost and offered the least risk (if the teacher can't teach CAD on them or until something better comes along, the micro can be used for something else like teaching word processing). The micros have helped introduce students to CAD; but, they have, until now, lacked the power and sophistication necessary to "train" CAD operators.

Simply stated, there are too many bells and whistles constantly being added to CAD equipment for most educational institutions to make an instructional "best buy" in terms of hardware and software. As one teacher recently commented, "This is a great CAD system, I just wish I knew how to make it do all the things it's supposed to." Plus, since most CAD systems tend to address one specific drafting field such as architectural, mechanical, or electronic over other fields, they present a problem for educational institutions who usually teach several different drafting fields. On a more positive note, slowly, but surely, the instructional basics of CAD essential to operational competence are being identified and appropriate instructional materials developed.

FACULTY COMPETENCE

No matter how good the equipment is, how much equipment you have, or how much courseware is available, it is the faculty member who determines how good the instruction will be. Teaching is one of the few professions which occurs behind closed doors and is rarely practiced in front of peers. Any administrator who has been around long enough to try forcing a change in the curriculum against a department's or division's wishes will quickly allow that no matter how good the argument or how strongly you make your feelings known, the faculty members will have their way.

The old adage is very true, "teachers teach as they were taught," and we have a generation of drafting teachers who were taught conventional drafting techniques by instructional methods which have been in place almost since the turn of the century. Schools and colleges have a massive retraining process on their hands if they intend to integrate CAD into the drafting curriculum. As was mentioned earlier, this is not simply a matter of transferring learned psycho-motor skills from a manual medium to an electronic medium as can be done with typing and word processing. In CAD, the primary tool being transferred is knowledge and not manual skills. Concepts, associations, configurations, and abstractions take on an even more important dimension in CAD as opposed to eye and hand dexterity.

COST OF INSTRUCTION

Drafting is, and has been, a technical occupation taught in a laboratory setting where each student is provided a drawing board or workstation. Those traditional drafting workstations seldom cost more than a few hundred dollars each. The dominant instructional cost was the salary paid the instructor and a few supplies. With CAD, a single student workstation can cost more than all the equipment combined in the traditional drafting lab plus the salary for the instructor and the supply budget.

Since CAD equipment remains an expensive item for most institutional budgets, the limitations imposed by the equipment chosen sets the instructional parameters in terms of what is taught (e.g., 3-D, programming applications) and how many students the institution can serve. There are few colleges and schools who can afford to spend $100,000, or more, each year on their drafting programs. Typically, an institution will seek special funding for a one-time purchase much like they would spend in setting up a word processing lab, a machine lab, or an administrative computing system. Once established, most institutions are forced to stay with the same configuration until it is "their time again" to retool their instructional program.

Colleges and schools are caught in a "catch-22" situation with regard to CAD equipment. If more CAD workstations are bought at a lower price, then more students can be taught the basic skills and concepts of drafting. But, they are not afforded the opportunity of advanced training or production-oriented laboratory exercises. However, if more expensive industrially-oriented CAD systems are purchased, then fewer CAD workstations will be available for student use; thus, fewer students will be afforded exposure or introduction to CAD. This dilemma largely results from the present, heavy equipment-dependent nature of CAD instruction when taught within the traditional fixed laboratory format as found in most drafting classes. This type of fixed format requires a set period of time for instruction, is strongly teacher-centered, and assumes a workstation for each student. For CAD instruction, it is usually suggested that no more than two persons be assigned to one workstation. Assuming establishment of a CAD lab similar to the traditional drafting lab, the equipment costs will be among the highest for all vocational-technical programs.

Given the preceding circumstances, implementation of the traditional drafting lab model for CAD training may be financially prohibitive for most educational institutions. As a result, important instructional concerns are largely influenced by financial constraints.

TRANSFERABILITY OF LEARNING FROM THE CLASSROOM TO THE WORKPLACE

Before discussing transferability of learning, it might be helpful to identify the primary educational institutions most likely to offer CAD education and their respective teaching responsibilities. At the university level, institutions teaching CAD address two separate groups of students, neither of which is destined to be CAD operators or technicians. One group of students is those majoring in industrial technology whose objectives are closely aligned with vocational-technical education. The other group of students is those seeking engineering degrees where CAD is taught as one of the supporting fields.

The secondary schools providing CAD training for students are limited but growing. Their major goal is student exposure to CAD in order for students to have a better appreciation of what CAD is and what it can and cannot do. Increasingly, however, there is growing public sentiment that vocational-technical education should not be the responsibility of the public schools. It is argued that the schools should concentrate their efforts toward general education and not skills training.

Clearly, postsecondary institutions at the two-year college level and area vocational-technical schools will carry the largest public burden of training persons as CAD operators and technicians. For these institutions, transferability of learned skills (motor and cognitive) from the classroom to the workplace is critical to their mission. One major problem they face is the determination of an acceptable level of competence for students completing CAD courses. There is yet to be defined a generally agreed-to set of competencies for CAD operators or technicians. Once defined, these student competencies will be a major determining factor for equipment purchase and curriculum design.

Central to the preceding issues and concerns is a reasoned approach to the identification of critical teaching issues and the development of instructional materials based on sound learning theory. The present CAD hardware and software, assuming use of raster and hard disk, are sufficiently advanced technologically to meet the instructional needs of most institutions involved in operator and technican training for the next few years. What is needed is a major commitment to the development of instructional materials designed to (1) provide an acceptable level of instruction in CAD applications in several different drafting fields, (2) provide student with an opportunity to learn input of data from the primary entry mechanisms--presently those are keyboard and menu, (3) reduce the present equipment-dependent instruction costs for CAD, and (4) provide drafting faculty with needed additional support in implementing CAD as a new tool. The next section addresses some of the key instructional design considerations and observations on CAD instruction at educational institutions.

3. INSTRUCTIONAL PLANNING PRECEPTS FOR CAD

Based on the preceding issues and concerns, some general instructional design guidelines and observations about CAD instruction are presented. These guidelines and observations on CAD instruction result from numerous meetings and discussions with CAD educators and instructional design specialists. They follow traditional educational precepts important to student learning and were used in the development of a CAD learning system designed to meet some of the issues identified in the previous section.

INSTRUCTIONAL GOALS AND DESIGN GUIDELINES

The following goals and guidelines address the preceding issues and concerns of educational institutions for CAD instruction. They are not meant to be inclusive but to serve as planning benchmarks in the development of a CAD learning system.

General Instructional Goals. These general goals address the preceding instructional issues with special focus given cost-of-instruction, transferability of learning, and "instructor-friendly" teaching materials which are based on sound learning theory.

(1) A CAD system (hardware and software) for educational institutions should offer the latest technology and be industry-standard in its operation so as to assure the highest possible dgree of relevance and transferability of learning to the workplace.

(2) Comprehensive courseware (e.g., manuals, aids) for teaching of CAD should be developed, with the help of drafting educators and instructional designers, which is based upon sound learning theory and use of the latest learning technologies for delivery of instruction.

(3) A CAD system for education, because of increased instructional costs, should be instructionally designed to permit economy of operation through such features as flexibility, self-paced learning, open-lab instructional format, and multi-course (introductory through advanced) capabilities.

Instructional Design Guidelines. Based on the preceding goals, these guidelines offer developers of instructional materials and learning systems some key points which should be considered in the development of future instructional materials for CAD.

(1) Courseware should be developed with input from persons actively engaged in the teaching of drafting and knowledgeable in instructional design.

(2) Instructional materials should be sufficiently diverse in method of presentation (e.g., video, print) to accommodate differences in student learning styles.

(3) Learning activities should be modularized into discrete units with clearly identified performance objectives related directly to the learning activities and evaluation procedures.

(4) The learning modules should be sufficiently flexible and independent to permit the instructor to customize a course(s) according to logically determined needs for CAD training.

(5) Courseware must be desired to maximize student motivation through such means as permitting access to equipment operation as early as possible, providing visual reinforcement for learning activities, and modularizing learning in discrete units of knowledge and skills easily mastered within a short period of time.

(6) Courseware should, as closely as possible, provide the student with "real-life" training materials he or she would encounter when employed as a CAD operator, technician, or supervisor.

(7) Since the teacher is the most important factor in the instructional process, the courseware should be "instructor-friendly" permitting the teacher an array of learning materials to choose from when configuring a CAD course or a CAD curriculum.

(8) The courseware should permit effective integration of CAD into the traditional drafting room as a tool rather than as a separate discipline.

(9) Courseware should be designed for use from the very introductory level classes through the training level.

(10) Courseware should be designed and presented in a format which will allow for easy inclusion of additional or revised instructional materials, minimizing the need for additional investments in equipment.

These goals and guidelines are not meant to be all-inclusive but to provide a benchmark for those vendors, publishers, and educational institutions involved in the development of CAD learning materials.

4. SUMMARY

Educational institutions have admireably responded to the need for training CAD operators and technicians. Their response had been hampered by the (1) lack of adequate instructional materials, (2) need for faculty retraining, (3) an acccelerating cost-of-instruction due to CAD hardware costs, and (4) unclear expectations as to realistically established student competencies and the transfer of learned CAD skills from the classroom to the workplace. These issues are now being raised at the state level with accountability being demanded for the expenditure of monies to purchase CAD equipment.

Of the preceding issues, the lack of adequate instructional materials to responsibly implement CAD training is considered, at present, the greatest need. It is unlikely that CAD vendors will address this problem since hardware and software is their prime concern. Publishers have generally avoided the need for CAD instructional materials. And educational institutions, themselves, have engaged in only limited sharing of instructional resources for CAD.

What is needed for CAD instruction is a comprehensive set of instructional material which will initially provide (1) drafing faculty with a highly structured learning package both for instructor use and student use and (2) sufficient instructional materials to permit faculty, as they become familiar with the system, to customize courses to meet both common and unique drafting needs (e.g., architectural, mechanical, electronic). To accomplish these goals, several general instructional goals and objectives were presented as precepts for developers and designers of instructional materials for CAD.

BIOGRAPHY

CECIL L. GROVES, Manager of Educational Applications for Houston Instrument, was formerlly college president of Austin Community College in Austin, Texas for 6 years and executive vice president of Delgado College in New Orleans for 7 years. He presently serves as Chairman of the Technical, Trade, and Industrial Committee of the National Computer Graphics Association. Dr. Groves received his Ph.D. from The University of Texas in Austin in 1970. He is a frequently published writer and a recognized speaker on applications of technology in education. As a former board member of the American Association of Community and Junior Colleges, Dr. Groves has served on numerous state, regional, and national committees concerning postsecondary education and training.

13

A STRUCTURAL MODEL OF CREATIVE PROCESS FOR IMPROVED INTERFACE DESIGN

Robin G. King
Sheridan Computer Graphics Laboratory
Sheridan College of Applied Arts and Technology
Trafalgar Road
Oakville, Ontario L6H 2L1 Canada

Abstract

This paper examines the development of a single structural model of creative process and explores the implications for graphics interface design. The need for an open systems framework for studying the dynamics of graphics interfaces is identified, followed by a brief review of traditional approaches to the study of creative behavior and innovative processes.

A single structural model of creative behavior in the form of a topological "map" is presented using catastrophe theory as a tool for the analysis of discontinuous behavior in open systems. Recommendations for specific empirical studies are suggested.

1. INTRODUCTION

The purpose of this paper is to present a topological model of creative behavior as a basis for, and enhancement to, the process of graphics interface design.

The most common and successful techniques for interface design take the form of the methodologies formulated and detailed by specialists who have concentrated on task analysis. While it is not the purpose of this paper to review these techniques, an alternate structural framework is described which may prove essential to the development of graphics and computer-aided-design interfaces which respond in a more natural way to the artist or system operator.

The position argued here is that the most effective methods of developing interactive graphics environments will not be maximized until the design system and environment are treated as a dynamically evolving open system, responding in appropriate temporal sequences to the dynamics of innovative change.

In order to realize this proposition, we must fully understand the nature of this system and the ways in which its dynamic interrelationships change over time. By using the techniques of general systems theory, the analysis of a typical CAD or computer graphics system might be formulated as shown in Figure 1.

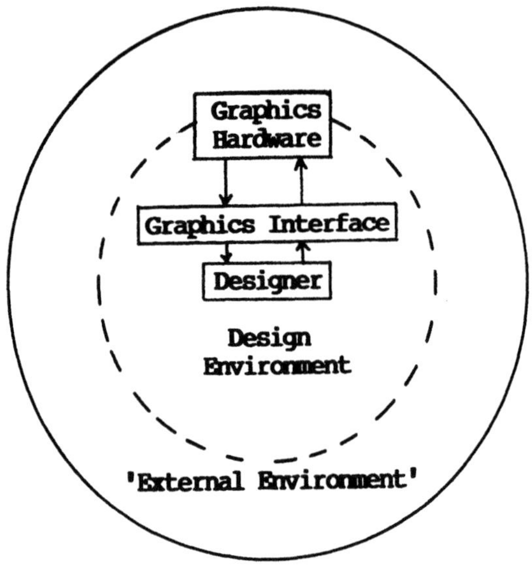

Figure 1. A General Systems Model of a Typical CAD/Graphics System

In simplistic terms, the system consists of a dynamic arrangement of graphics hardware, graphics interface, designer, physical environment, and design environment. The last of these, the design environment, must be considered carefully. The design environment is seen here as a "mental construct" which the designer holds as whole or complete at a particular time during the creative design cycle. It may be regarded as a sub-system, the boundary of which changes in both nature and permeability and therefore has a temporal rather than static quality.

During the different stages or phases of creative activity, the nature of this boundary changes in response to the designer's mental and physical state, in particular to the psychological states which characterize innovative activity. This "conceptual sub-system" is illustrated in Figure 2.

It is evident from this illustration that the forces at work during the design cycle--and throughout innovative and creative processes--are much more complex than it would first appear. Not only is there information flow related to the specific structure, design, and dynamics of the graphics interface, but the effects of emotional, intuitive, and intellectual forces have considerable influence on the type, range, and nature of changes in the system's behavior.

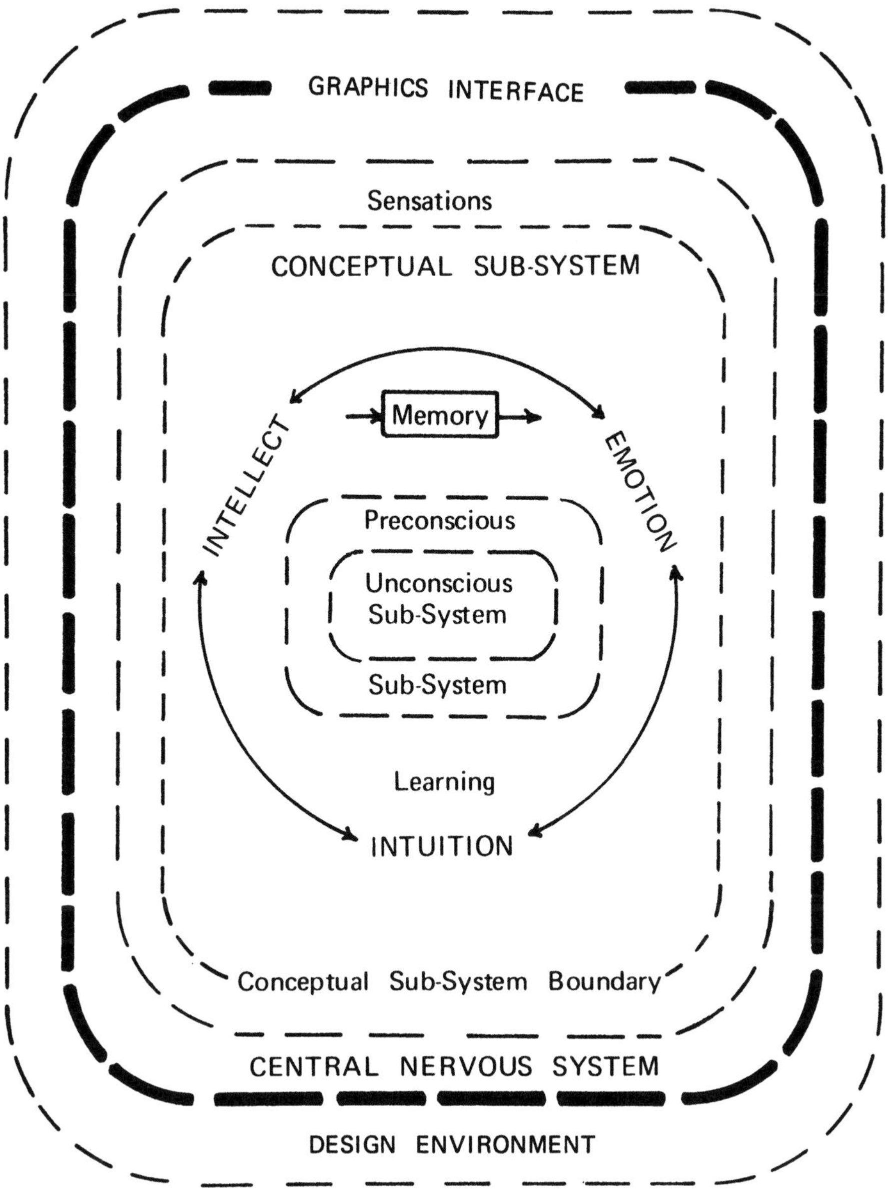

Figure 2. The Design Environment

We will limit this discussion to the systems dynamics during creative behavior in order to show that a more thorough understanding of creativity must be a central concern to the graphics interface designer. Graphics designers (and, incidentally, all others who use these systems--animators, illustrators, film makers, programmers, etc.) employ many different modes of thinking during their time at a terminal and are by no means limited to structured problem solving techniques.

2. THE CREATIVE PROCESS

Creative process may be defined as "the transformation or emergence of new ideas or artifacts which result from a series of interconnected actions or events." Wallas (1) was the first to propose a division of the creative process into a series of more-or-less discrete phases or stages. These he termed preparation, incubation, illumination, and verification and these terms are still commonly employed in the field of creative studies. Vinache (2) has been particularly critical of the idea of discrete stages and quotes Wertheimer's argument that:

> The real weakness in the view that creative thinking consists of fairly well-defined phases is not that these stages do not exist, but that they are regarded as universal, successive and distinct from one another. In actuality it would be better to conceive of creative thinking in more holistic terms, as a total pattern of various processes which overlap and interweave between the occurrence of the original stimulus and the formation of the final product. (Wertheimer, as quoted in Vinacke (2))

A more complete and satisfactory way of looking at creative process is to regard it as a type of "whole field phenomenon". Imagine the four phases of creativity as the corners of a pyramid. We may then treat the process as a whole by regarding it as a shift of <u>emphasis</u> of activity among the four phases in any direction. The leads us to a more complete and holistic picture which recognizes the immense variety to be found in approaches to creativity by different individuals. Figure 3 illustrates this clearly.

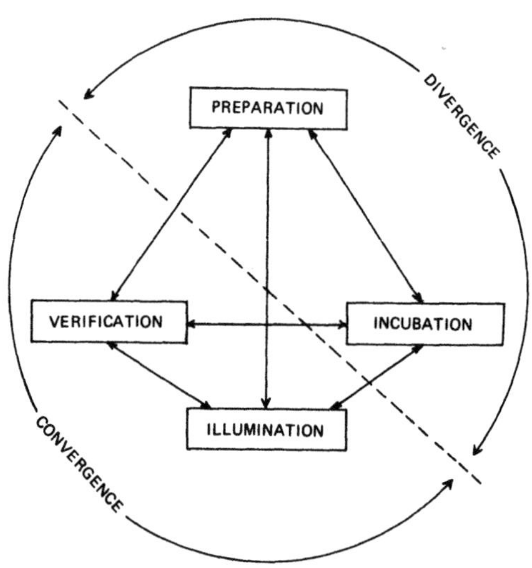

Figure 3. Creativity as a "Whole Field Phenomenon"

Note that the system is "contained" in a field of mental activity divided into two regions or principle <u>forms of thinking</u>, convergent and divergent. The ability of the creative individual to use both types with equal and combinatory dexterity is essential to the development of original thinking, as each stage of the process is attended by a particular set of mental conditions. The psychological patterns which accompany each phase are approximately as follows:

2.1 PREPARATION

The preparation phase is marked by highly divergent mental activity; the person searches for variety of stimulus in the environment. The individual explores his or her surroundings, looking for stimulus, or is presented with or discovers a problem which merits attention and appears worthy of further investigation. The initial impetus for creative activity and for the preparative phase in particular comes during periods of psychological openness.

This openness appears to be crucial for any significant endeavor. The individual may, for example, attend lectures, read, listen to music, exercise, contemplate works of art, engage in lengthy discourse, drink, make dream records, or meditate. Whichever method is used, exploration is the key, enabling that person to investigate his or her intellectual, emotional and intuitive feelings and reactions through contemplation and inquiry. Not only is the creative person reaching out into the environment for stimulus, but the journey is internally oriented, searching the realms of the unconscious as a means of initiating the innovative impulse.

During this phase the design system's boundaries tend to be highly permeable, allowing for large amounts of information to flow into or out of the system and among its sub-systems. The result of the highly divergent, open, discovery-oriented phase of preparation is a movement towards a more convergent activity. The individual recognizes (either intellectually or intuitively) that sufficient divergence has taken place and then settles into a phase of either convergent attempts to put the information together or--more likely--into a phase of relatively inactive conscious effort. The problem appears to be driven "underground" into the unconscious where it is manipulated without apparent conscious control by the unpredictable forces of association. This incubation period seems to function as a time of internal elaboration, reorganization, and synthesis of new solutions to the problem. The mechanism by which this happens is not well understood!

2.2 ILLUMINATION

The products of unconscious incubation usually reappear as a sudden "inspirational" leap of association in which the new or novel solution surfaces to consciousness. This may happen shortly after the conscious work on a problem has started or it may surface after years of incubational activity. The release or appearance of the information usually happens when the person is relaxed and does not appear to be of the individual's volition (except that induced states of relaxation tend to aid the process).

2.3 VERIFICATION

Spectacular as the sudden processes of illumination often are, the solution often surfaces only as an image or intuitive feeling, although there are accounts of fully-formed ideas--especially in the form of scientific and mathematical solutions. The creative person must at this point, either individually or as a part of some collaborative team effort, work through a number of convergent associations or connections in order to make the insight a practical working proposition. This phase of the activity generally consists of difficult convergent, synthetic effort, often accompanied by much trial and effort.

3. CREATIVE BEHAVIOR

The study of creative behavior and the development of descriptions and explanations of creative process have traditionally been divided into four distinct but interdependent areas of exploration:

(1) Descriptions and theories relating to the behavioral characteristics of the CREATIVE PERSON.

(2) The study of the particular properties and forms of the CREATIVE PRODUCT.

(3) The influence and nature of the ENVIRONMENT as a contributory factor in creative activity.

(4) The various stages and phases of the activity by which a new or novel artifact is developed, the PROCESS.

This paper offers a practical alternative to the reductionist approaches of the past by offering instead a holistic, single structural framework within which these aspects are combined but which, at the same time, respects their variety and contributions. Before this framework can be described, a brief description of the above orientations is in order.

3.1 THE PERSON

The personality of creative individuals (and indeed the personnel climate of a corporation) is probably the single most important factor affecting the outcome of creative endeavor. An individual's beliefs, interests, ideas, values, impulses, opinions, working methods, and physical and mental attitudes are all summed up in the essential quality we call personality. Many theoretical approaches and the most significant part of research dealing with creative people focuses on the personal attributes which help or hinder creative activity. Four different approaches cover most of the available research.

3.1.1 Literary Accounts

Literary accounts of creative behavior are centered on biographical and autobiographical records, both idiosyncratic and comparative. Descriptions and writings by famous individuals are numerous in both the fine arts and science, frequently concentrating on the unusual behavioral nonconformity and personal idiosyncrasies of creative people. While the accounts most frequently cited relate to the lives and working methods of artists, there are numerous detailed studies which explore the lives, habits, working methods, and creative behavior of scientists, philosophers, historians, and other innovators.

While no single picture emerges from the literature, the important contribution of literary writings is the immense variety and richness of human creative endeavors, covering the complete spectrum from fully-functioning individuals to those gripped by apparent psycholgoical and physical illness. The essential notion to be gained from this is that the design of creative interfaces must respect the richness and variety of different creative processes and design methodologies.

3.1.2 Psychological Theories

Many theories have been proposed to account for the mechanisms of creative thought, motivation and behavior. Among the best known are those proposed by Freud, Kris, Kubie and others. In recent years, Koestler (3) has contributed the idea that crucial to the creative act is "the ability of the individual to bring together two or more diverse or apparently ambiguous ideas or images."

Among the humanistic psychologists, Maslow's (4) concept of "openness to experience" and Roger's (5) requirement of "an internal locus of evaluation" (compare with design environment) both appear critical themes for the selection of appropriate personnel and for development of working relationships. Some writings have surfaced which emphasize a developmental perspective for the enhancement of creative behavior, stressing the improvement of creative thinking over time through personal growth and training. At the same time, we have seen the concepts of "split-brain" and holographic models of creative thought surface to some prominence.

3.1.3 Clinical Studies

Clinical studies result from the observations made of creative individuals who have undergone psychological assessment or treatment. They are generally centered on some aspect of the individual's neurological problems or psychological abnormalities. The overall contribution of these studies is at best questionable because the majority of creative individuals seldom exhibit abnormal psychological patterns in their behavior.

3.1.4 Psychometric Studies

Psychological research attempts to identify through testing procedures the specific patterns of psychological traits or characteristics which are common to creative individuals. This research centers on the commonalities and differences evident between the various sub-groups of the population such as artists, scientists, musicians, the gifted, and various cultural and ethnological groups. Some of the characteristics apparently common to creative people include: higher than normal spontaneity, intelligence, fluency and flexibility of idea production, originality, scepticism and curiosity.

During the past twenty-five years, a number of researchers and organizations have developed systematic and structured procedures for the development of individual creative capabilities and corporate innovation. Among the former, the Creative Education Foundation of the State University of New York at Buffalo has researched and implemented a number of strategies for structured problem solving and has developed techniques for enhancing both the quantity and quality of potential ideas. Many companies have developed their own corporate systems for innovation and others, such as Synectics Inc. of Boston, have surfaced to provide systems for enhancing creative environments, communications, and innovation techniques for the corporate environment.

3.2 THE CREATIVE PRODUCT

The creative product must be, by definition, essentially new or novel before we can consider other criteria by which to measure its creative characteristics. While the definition of novelty is a difficult one, we shall assume for this paper that a creative product is one characterized by an emergent property; that is, something has surfaced as part of the product or idea which is unique. The majority of writings on this subject can be found in the descriptions, aesthetic and stylistic criteria by which different societies judge works of art. These are usually based on form, content, criticism, and aesthetic preferences.

The products of scientific inquiry have not been so frequently addressed but that the phenomena of novelty and creative breakthrough have been the subject of philosophers of science. Notable in this field is the contribution of Thomas Kuhn (6) whose work on the nature of shifting scientific paradigms has done much to spark assessments of changing thought in contemporary science. As the impact of scientific creative production has such an impact on our daily personal and business lives, the changes underlying basic thought in science are crucial to an understanding of the marketplace and our place in its dynamics.

3.3 THE CREATIVE ENVIRONMENT

There appear to be two opposing views regarding the environmental factors which promote or inhibit creative functioning. Some writers have stressed that a setting which facilitates a respect and explicit valuing of ideas, providing greater opportunity for self-initiated action, and a climate which reduces the fear of negative evaluation, will be one in which creative productivity is more likely to take place. Others have argued that creativity will occur regardless of negative environmental conditions and, in fact, it would appear that for some, the more chaotic the psychological and physical environment, the greater the productivity (this tends to be generally more true for artists than for scientists).

Having established the general patterns by which creative process takes place and briefly reviewed the various reductionist attempts to describe the characteristic attributes of creative activity, we are now is a position to bring these diverse ideas into a single structural model. The description of a systems model of creative individuals has been given elsewhere and in more detail (7).

4. A TOPOLOGICAL MODEL OF CREATIVE PROCESS

4.1 CATASTROPHE THEORY

Catastrophe Theory deals with the mathematics and dynamics of divergent and discontinuous phenomena. As a general systems model, it is both versatile and comprehensive, being structured around a set of universal principles which have been extensively applied to such diverse fields as animal behavior and physics by Zeeman (8), biology and evolution by Thom (9), sociology, chemistry and politics by Woodcock and Davies (10), health, psychology and metaphysics by Postle (11), and to creative process by King (12).

Catastrophe Theory was developed by Rene Thom, whose classic work on the subject, Structural Stability and Morphogenesis, was first published in 1972. Thom's thesis is centered on the dynamic mapping of the morphological changes in biological structures by the use of topological surfaces. The theory deals with both the nature and dynamics of continuous and discontinuous changes in systems behavior and is ideally suited to an analysis of the creative process which, by its nature, is characterized by a discontinuity, the phenomenon of illumination or intuitive insight. The rigorous mathematical foundations of the theory are not used in this particular application.

A catastrophe is "any discontinuous transition that occurs when a system can have more than one stable state or follow more than one stable pathway" (10). A generalized cusp catastrophe is illustrated in Figure 4, this particular form being one of the seven elementary or archetypal catastrophes which Thom has identified in his work and the one which seems most appropriate to the phenomenon under consideration.

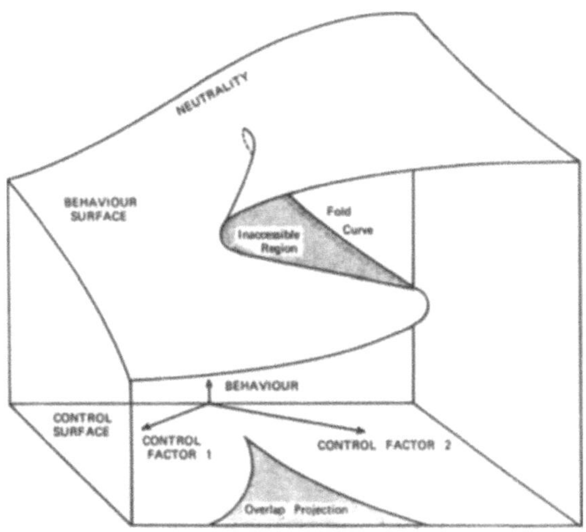

Figure 4. A Generalized Cusp Catastrophe

Two control parameters of the system under consideration are plotted on a horizontal plane called the CONTROL SURFACE. The behavioral component is then illustrated dimensionally on a third axis perpendicular to the control plane. If the range of behaviors for variations in the two control parameters are plotted, a topological surface termed the BEHAVIOR SURFACE, of the general form shown as a curved surface with a pleat, will result. The pleat is a result of the discontinuity of the phenomenon being described. Woodcock and Davies illustrate this as follows:

> The unfoldings are called catastrophes because each has regions where a dynamic system can jump suddenly from one state to another, although the factors controlling the process change continuously (10:p25)

Five important properties of a cusp catastrophe have been identified viz., bimodality, divergence, sudden transitions, hysteresis and inaccessibility. These forms will be described in general terms, followed by a specific application to the analysis of creative process.

Consider two conditions of a system with equivalent neutral behavior at a point A on the Behavior Surface. If these two conditions, under equal influence by the control factors, move across the surface, they will parallel one another until they reach the start of the fold line. At this point, one may move across the upper surface of the pleat and the other across the lower surface (as shown in Figure 5).

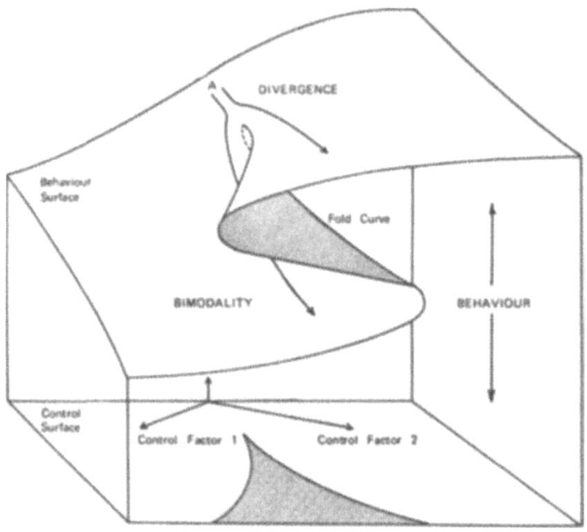

Figure 5. Bimodality and Divergence

The behavior of the two conditions of the system is similar or the same at the outset but markedly different in their respective new positions. A system which exhibits this type of behavior is said to show DIVERGENCE. It can be seen from the above that a system may therefore show two possible types of behavior for a specific combination of control factors. When this is so, the system is said to be BIMODAL. As the two control factors change (as the systems behavior changes), the system in question may make a variety of SMOOTH TRANSITIONS or changes as indicated in Figure 6.

But not all systems transitions are smooth. Consider a system at a point Y on the semi-stable fold line at the bottom of the pleat. For an increase in control factor 2, the only behavioral change possible is a sudden jump or transition to the upper surface of the pleat, to a different type of behavior. Similarly, a system at a point Z may make a change through a similar transition. These types of behavior are termed SUDDEN TRANSITIONS.

The behavior described is characterized by a sudden discontinuity. The underside of the pleat, between the top and bottom surfaces, is generally considered to be INACCESSIBLE. Small changes or slight disturbances in the relative amounts of the two control factors ensure sudden transition to one or the other of the two behavioral surfaces.

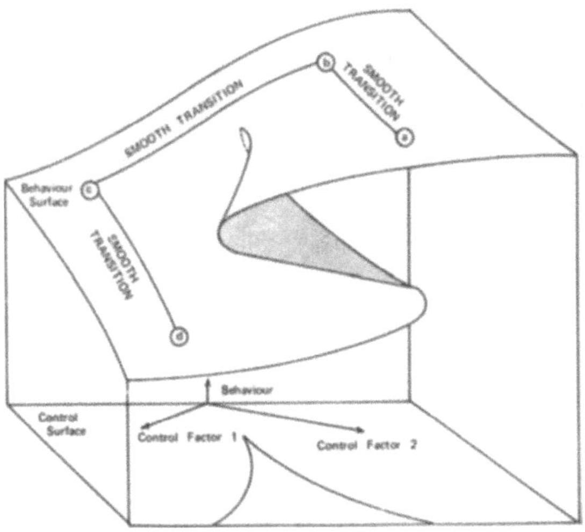

Figure 6. Smooth Transitions

The last characteristic is termed HYSTERESIS. A cycle of changes which includes two sudden transitions alternated with smooth transitions is said to show hysteresis, as shown in Figure 7.

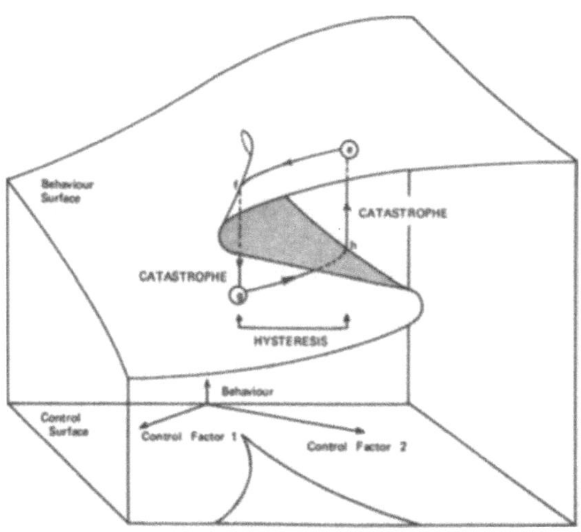

Figure 7. Hysteresis in a Cusp Catastrophe

Given these five properties of a cusp catastrophe, we may now describe their application to an analysis of creative process, the development of a single structural model and the implications for graphics interface design may be explored.

4.1.1 The Creative Process

A creative behavioral topology may be proposed in the form illustrated in Figure 8. Two control parameters have been chosen, convergent and divergent thinking, these having been identified as characteristic of creative activity. The four phases of creative process, i.e., preparation, incubation, illumination, and verification, are shown in their respective positions governed by the relative amounts of convergent/divergent thinking evident during each phase.

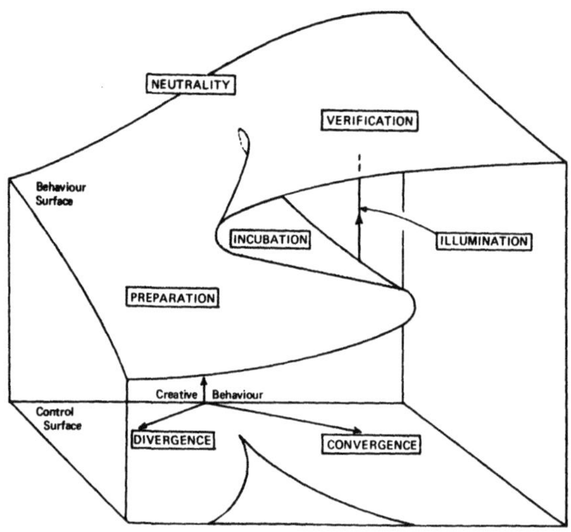

Figure 8. The Creative Process Topology

Preparation, characterized by highly divergent thinking, is shown as the bottom surface. Incubation, the relatively inaccessible phase, appears as the behavioral characteristic of that portion between the two surfaces. Illumination is shown as the convergent factor which causes the system to move from divergent search to convergent association (the sudden transition) and verification appears as the upper surface of the topology. Given this initial structure, we may now examine the possible dynamic conditions of creative process in terms of a single catastrophic model.

4.1.2 Smooth Transitions

Creative process as we have shown generally starts with divergent search. The human creative system moves from a point of relative "neutrality" as far as the problem is concerned, to the lower portion of the behavioral surface.

During divergent search, frequent smooth transitions may be made as the individual explores the experimental environment and attempts to concretize possible solutions by varying amounts of convergent thinking.

In addition, as part of the divergent activity, frequent movements into the "inaccessible" incubational region may occur, but the divergent dynamic provides a stronger stimulus than the convergent, the system tending to return to the lower part of the behavioral surface. This dynamic may be repeated many times until the system becomes "saturated" with the information, facts, or images relating to the problem. At this point, the system is ready for a "creative catastrophe".

4.1.3 Sudden Transitions

The creative system, saturated with divergent information, will naturally attempt convergence, will try to put this information together and move across the lower portion towards the pleat as further convergence is attempted. The system will then be located at a point on the semi-stable point of inflection on the fold line of the pleat. Further movement towards convergence is then only possible through a process of sudden transition to the upper surface.

This sudden transition is the illumination phase of creative process. The system's behavior makes a dynamic jump, a quantum leap, after a period of incubation in the inaccessible region. While there is virtually no change in the relative amounts of divergent and convergent control factors, the incubational activity provides the "energy" needed to transform the system through a quantitative change to a convergent dynamic condition.

One has only to complete the circle of points to form the traditional cyclic model of continuous and discontinuous phase changes which characterize the creative process, as shown in Figure 9.

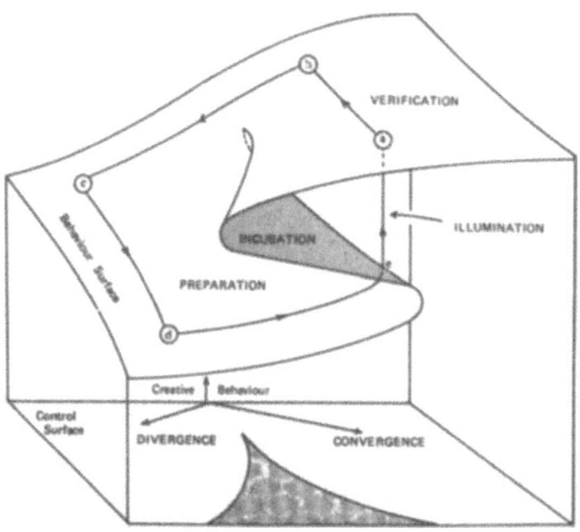

Figure 9. Continuous and Discontinuous Phase Changes

4.1.4 Creative Breakups

An alternative catastrophe, a jump from the upper surface to the bottom surface, may be observed. This is a time when, having succeeded in the associational task, the images or ideas appear incomplete and additional attempts to incorporate a wider perceptual field cause a reversed catastrophe --everything falls apart. This broadening of the situation pushes the system into divergence, requiring further creative effort.

This TRANSFORMATIONAL JUMP is really a breakthrough or breakup (even though it looks down on the model) to fresh perceptions on the problem and moves the system towards further divergent search. This process, and the conditions under which it happens, have been virtually ignored by researchers in this area, and yet it may be almost as important as the creative illuminational experience itself. In fact, many artists recognize this and build into their creative processes chaotic conditions which tend to promote disorder (13). This cycle of double sudden transitions has the characteristic of hysteresis as illustrated in Figure 10.

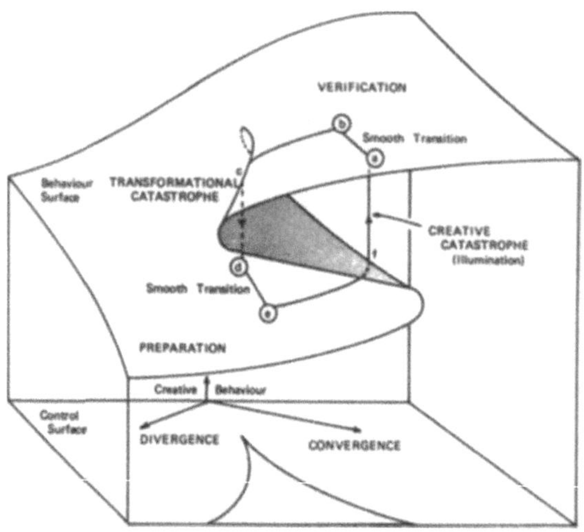

Figure 10. Double Catastrophes in the Creative Process

The holistic general systems model presented here is unique in several ways:

(1) It provides a SINGLE STRUCTURAL FRAMEWORK within which the complete range of possible sequences of behavior constituting creative activity may be viewed.

(2) The EVOLUTIONARY PERSPECTIVE of the model is evident through the emergence of new perspectives (creative breakthroughs) which result from fluctuations in the system's states (14).

(3) The INFINITE VARIETY of different creative processes can be viewed simultaneously.

(4) It is COMPREHENSIVE because a wide variety of existing theories and observations about creative phenomena can be subsumed within one model.

5. THE IMPLICATIONS FOR GRAPHICS INTERFACE DESIGN

Computer aided design systems have already become a major force in the design field. With the mass production of high-resolution computer graphics systems expected in the near future, and the continuing advances in image reproduction technology, advances in the quality of graphic interface design are critical. Empirical research into the relationship of the technology to creative activity is desperately needed. Some specific suggestions for research are given below.

First, an investigation should be made of the psychological changes which take place during and after adoption of these technologies. A systems-oriented approach to user characteristics and behavioral changes needs to be completed. Psychological measuring techniques which may be used for this purpose are unfortunately rare. Luckily, one system of psychology, George Kelly's Psychology of Personal Constructs (15), holds great promise because it conceptualizes an ideal structure within which to study human-computer interfaces.

Kelly's fundamental postulate, "a person's processes are psychologically channelized by the ways in which he anticipates events", provides the key to his system. While his structural framework will not be described in detail here, it is suggested that the Repertory Grid Technique developed by Kelly is an ideal tool for the above analysis. It is based on the identification of an individual's system of "constructs", (referring to the templates or patterns used by an individual to view the world). Previous work by the author (1) has demonstrated both the feasibility and utility of this tool for analyzing the evolution of construct systems resulting from changes in expertise within visual artists.

Proposed research activities should include:

(1) Studies of personality differences and their relationship to creative productivity with different systems and designs.

(2) Identification of cultural differences, particularly with respect to attitudes, perceptions and differences of process.

(3) Longitudinal studies of personality or attitudinal change resulting from interaction with this technology.

(4) Longitudinal studies of change in creative behavior patterns and techniques.

(5) Identification of the psychological blocks to creative thinking resulting from specific interface designs.

(6) The study of long term environmental influences on design process and methodology.

(7) The development of testing procedures for new interface designs.

(8) The development of a method of comparison for relating traditional graphic design methodologies with those used in interface design.

As more evidence of the positive and negative effects of divergent interface designs surfaces, the productivity and appropriateness of this technology to the solution of problems requiring high degrees of innovation will demand that much of the above research will be required. The quality of interface design has already been found to significantly affect the design and structural frameworks used to develop education and training programs (16).

6. CONCLUSION

This paper has demonstrated the advantages of a general, open systems model of creative behavior as an analytical tool for empirical research into graphics interface design. It has been suggested that such a tool fills a critical vacuum in interface design methodology, and that research using Kelly's Repertory Grid Technique could aid in designing interfaces which respond more appropriately to both the individual differences, and commonalities, which characterize the creative designer.

7. REFERENCES

1. Wallas, G., The Art of Thought. Harcourt, Brace, New York, 1926.

2. Vinache, E.W., The Psychology of Thinking. McGraw-Hill, New York, 1974.

3. Koestler, A, The Act of Creation. Hutchenson and Co., Great Britain, 1964.

4. Maslow, Abraham, Toward a Psychology of Being. Scribners, New York, 1968.

5. Rogers, Carl, On Becoming a Person. Houghton Mifflin, Boston, 1961.

6. Kuhn, Thomas S., The Structure of Scientific Revolutions. University of Chicago Press, Chicago, 1962.

7. King, Robin G., "The Application of General Systems Theory to an Analysis of the Creative Process in Visual Art. Unpublished M.A. Thesis, 1979.

8. Zeeman, E.C., "Catastrophe Theory", Scientific American 134, #4, pp. 65-84.

9. Thom, Rene, Structural Stability and Morphogenesis: An Outline of a Theory of Models. Benjamin, Reading, Mass., 1975.

10. Woodcock, A. and Davies, M., Catastrophe Theory. Dutton, New York, 1979.

11. Postle, D., Catastrophe Theory. Fontana, Frome, Somerset, 1980.

12. King, Robin G., "A General Systems Theory of the Creative Process", Proceedings of the Twenty-Fifth Annual Meeting of the Society of General Systems Research with the American Association for the Advancement of Science, pp. 121-130.

13. Peckham, Morse, Man's Rage for Chaos. Schocken, New York, 1967.

14. Prigogine, I., Thermodynamics of Irreversible Processes. Wiley-Interscience, New York, 1967.

15. Kelly, George, The Psychology of Personal Constructs. W.W. Norton, New York, 1955.

16. King, Robin G., "A Graduate Level Program in Computer Graphics for the Visual Arts", Proceedings of the Conference 'Computers in Art, Design, Research and Education', San Jose State University, San Jose, 1984, pp. 105-110.

8. BIOGRAPHY

Robin King started his career as a Research Technologist in the Physics Department of the Royal Military College of Science in England, and continued at the University of Waterloo after immigrating to Canada in 1965. For ten years he taught and coordinated the Applied Photograph Program at Sheridan College in Ontario.

Mr. King graduated with an M.A. from York University in 1979, specializing in the application of general systems theory to creative process in the visual arts and in 1982 was appointed a Colleague of the Creative Education Foundation at S.U.N.Y. Buffalo. He is currently the Coordinator of Sheridan Computer Graphics Laboratory and Education Director of the National Computer Graphics Association (Canada).

14

NAVY CAD/CAM

Captain John F. Leahy, III, USN
Program Manager, Information Resources Management Improvement
Naval Sea Systems Command
Washington, D.C. 20362

The assumptions underlying the following discussion are that CAD/CAM must be implemented in distributed systems, and that the data stored should embrace not only engineering data but also production planning, material and production scheduling, purchasing and material control, material handling, instructions to machine tools, production management, quality control, and management of these functions.

General Motors has a good grasp of this and of the needed universality of automated applications, and of the potential and scope of CAD/CAM installations in the factory of the future. When groundbreaking ceremonies were held for a new GM automated factory recently, a robot was present to place a pail near the selected site, lift a shovelful of earth, place it in the pail, pick up the pail, and return to work.

The primary importance of automation technology and the liklihood that it will become the new standard may be cause for concern by labor, but should be cause for alarm by management, for the following reasons. There will probably not be a drastic effect on labor as this technology is introduced into conventional plants. In 1820, 71 percent of the U.S. work force was engaged in agriculture. Today, agriculture employs 3.3 percent of the work force, and Americans are much better fed. The United States absorbed a shift of 68 percent of the population out of agriculture. U.S. manufacturing has never employed more than 20 percent of the work force, a level reached in 1920 and which today is 19 percent. Approximately one third of those people work in industries that already have heavily mechanized process systems, such as one sees in a chemical process plant or in what once was called Detroit-style automation. If all the most optimistic projections about CAD/CAM come to pass, the fraction of the work force employed in manufacturing may fall to perhaps 7 percent in 50 to 100 years. All the people who will be displaced will add to the new jobs that must be created elsewhere during the next 50 years by approximately 10 percent, or 2 percent per decade. That is much less than cataclysmic impact on labor. The major effect will be positive, in the form of an abundance of high-quality manufactured goods at low prices, with the result of an increased standard of living.

The impact will be greater on the professional class that replaces design engineers. Just as it now takes more education to operate a lathe than to handle a shovel, it will take still more education to be at the controls of an automated factory than to handle a drafting table or an on-line CAD system. The new and vital need to prepare engineers for CAD/CAM will require much more than learning FORTRAN or the operation of a CAD/CAM workstation. The educational challenge will be far greater. For example, two of the Navy's future aims are to build computer programs to execute design and planning, and to eliminate the need for NC programmers and supervisors on the factory floor who interpret the output of design and planning, that is, to feed the output of the CAD/CAM workstations directly to robots and NC tools. This implies that the people operating the CAD workstations and the software available to them will have in-depth professional knowledge and experience in design, planning, tooling, manufacturing, and quality control. By present training methods, such people graduate through 20 to 30 years of design and production experience in manual production operations, operations which may no longer exist. There are pressing unmet education and training meeds, and great opportunities for new training efforts in the CAD/CAM revolution, such as a training simulator program to teach new super-engineers the operation of an automated factory. Those who create and provide this urgently needed sevice are likely to profit from it greatly. Whether these new professionals, however they are eventually trained, will be called manufacturing engineers or automated factory controllers or some other title yet to be invented, U.S. engineering schools have a critical challenge to confront.

U.S. industry itself faces a much more immediate and profound challenge in the CAD/CAM revolution, and its response to the challenge will help determine the ability of the U.S. industrial plant to compete and survive in the coming era of automation technology, an era which for all of its progress has barely begun.

It has been said by some that factory automation cannot really succeed until expert systems are developed. Perhaps the Japanese, who are building Buick City and the Saturn car plants, and who already have achieved 25 to 85 percent cost and time improvements, are not aware of this. The potential value of artificial intelligence and expert systems is surely great, but the United States cannot afford to await their arrival. The Japanese are building 60 percent of all robots, while the United States and Western Europe together are building 40 percent. To maintain American industrial capacity in relation to that of Japan, U.S. production of robots should increase nearly sixfold, immediately. Government programs will not meet that need. The automobile industry, with both the motivation and the investment capital required, has made a commitment on the needed scale. Yet it appears to be alone in its efforts among industries with the potential to benefit from automation, nearly all others of which seem unaware of the urgency of the challenges to be faced. The urgency with which these issues are being discussed by U.S. industry and government leaders is far less than that required. Experts in automation technology should understand the critical need and potential benefits, but they do not seem to be educating others. The threat is clear and real: by failing to participate as a leader in the CAD/CAM revolution, the United States may forfeit its predominance and become a second-rate industrial power.

In the particular case under discussion here, the effort by the Navy to automate various phases of its shipbuilding manufacturing processes by CAD/CAM, experts were consulted who had developed CAD/CAM and industrial information systems elsewhere on a large scale. In brief, the consultants explained that:

(1) Expenditures of several billion dollars may be required in support of the attempt to automate the engineering and manufacturing process, and the attempt may fail. A system might be begun which could not be completed, or which could not be maintained if completed, or not operated if it were completed and maintained. The odds of failure can be great if the process of automation is not approached in a highly disciplined manner. There will not be sufficient funds to try twice. Before development begins, one must know exactly where, how, and when the payoff is to come at each step.

(2) The initial requirement is to secure the funds necessary to begin. (In the Navy case, this step is now under way.)

(3) Next is the requirement for good, comprehensive standards for all items of systems software and equipment and for all steps in system definition, development, implementation and evaluation. Of equal importance here is the required decision to strictly enforce these standards. One must be willing to fire people who attempt end-runs.

(4) Next, there must be a data base design for the whole process.

(5) Islands of automation must be avoided--islands separated by long queues of incomplete work. Divisions which are interested in automating only their areas must be required to cooperate in the program at large in the attempt to reduce cost and production time, and to improve quality through the entire process. One needs to learn to think in terms of the entire production process, which is a challenge by itself.

(6) Once a plan is made, one needs to follow the plan one increment at a time, earning a profit at each step so that the investment of additional funds can be justified. It must also be demonstrated that CAD/CAM can succeed in (the Navy's context) a government-operated industrial environment.

When asked what payoff would result, the consultants replied that the program eventually proposed would reduce the cost of Navy material by half, but they suggested that more conservative projections be made since few would believe such high figures. A cost reduction of 20 percent was said to be nearly guaranteed. Eighteen months have passed since that initial consultation, and most suggestions have been verified. The educational process within the Navy has begun, in a more extended fashion than might otherwise be necessary because of the need to carry out the process through the DOD. Nearly every group with something to contribute has been consulted. Current Navy CAD/CAM efforts have been reviewed, and these have been found to be succeeding in a local sense, but not in a global one. GM's purchase of EDS (for $2.5 billion) is understandable in this connection.

C. J. Date in his first book on database system presented the concept that there are three levels of concern in relation to databases. There is the level of users' views, or users' inputs and outputs. At a second level, the views must be related to a single, unified, logical database structure which is the concern of the database designer. And at a third level there is the physical system itself which consists of the stored data, access methods, and means of converting the physical data through the designer's logical structure to views, and vice versa.

It seems that users with too little knowledge of computer science and information systems development often are trying to operate on the hearts and brains of industrial systems without hiring experts or surgeons, and that they or the patients are dying as a result. Persons with knowledge only of the views are ignoring the expertise required to get the real engines designed and operating. This notion can be exteded to the problem of going from on-line graphic pictures to operating a computer integrated manufacturing system. It is not sufficient to put in a drawing and get out a drawing. The substantive knowledge in the drawings and the implications they entail must be converted to purchase orders, material and production schedules, instructions for tool development, instructions for material handling, instructions for production tools, and so forth. To accomplish this, it is necessary to involve users in defining the views, as well as highly experienced information system builders and database designers in defining the designer's logical data structure and the physical system.

It is clear that the building of a large CAD/CAM system is not something that can be addressed piecemeal and muddled through. Such systems must be developed incrementally, because they are simply too large to build as a single entity. Like naval ships, they are too expensive to build prototypes of.

As mentioned, the Navy is at the first stage of building a large CAD/CAM system, and the information about later stages reported above comes from others with more experience in this process. The problem of convincing the government to fund the program is still being addressed. Navy funds are allocated by people with the responsibility for being ready at all times to fight a war and thereby prevent potential enemies of the United States from thinking that starting a war against our interests would gain them more than they might lose. The Pentagon understands that military and naval planning must be based upon forces in being, not assessments of the other side's intentions. The other side can change its intention overnight. The composition of the fleet at any instant depends upon what we have done during the past 25 years and cannot be changed substantially in less than 10 years. People who allocate Navy funds properly are much more interested in buying naval ships and aircraft than in purchasing manufacturing process improvements. There is however a sufficiently large fund for improving the industrial side of the Navy if a substantial gain is achievable.

In attempting to initiate its CAD/CAM program, the Navy has passed the stage where people seriously questioned the premise that material costs could be reduced by 20 percent. It has reached a point where the Assistant Secretary of the Navy for Financial Management, who is responsible for all Navy computer-based systems, and the assistant Secretary of the Navy for Shipbuilding and Logistics, are verifying through external sources the information presented by Navy evaluators to them, examining CAD/CAM systems and results, and reviewing a proposal to establish a large system.

The segment of the Navy responsible for acquisition and logistic support of Navy material reports to the Chief of Naval Material and is called the Naval Material Command. Its five major segments are the Air, Sea, Electronics and Supply Systems Commands, and the Facilities Engineering Command. They are responsible respectively for naval aircraft, naval ships, electronic systems, supply systems, and shore facilities. The one responsible for ships, called NAVSEA, is the largest systems command. NAVSEA field activities stretch from Hawaii to Spain, its programs involve 111,000 military and (mostly) civilian personnel, and it manages 39% of the FY 83 Navy budget.

The title of the evolving CAD/CAM program is: Navy Integrated Computer Aided Design, Manufacturing, and Maintenance (NICADMM, pronounced Nick Adam) Program. Originally funds were requested for NICADMM for providing (1) overall program direction and standards, and (2) central procurement of standardized equipment and systems software plus the development of NAVSEA application systems. The intent was that other systems commands would independently request and justify funds for their applications. These commands asked to join their applications to the basic program, however, and the Navy Secretariat determined that such an arrangement would be desirable. Ways to centralize this effort are being evolved.

The National Academy of Engineering advised Congress in 1977 that exploitation of CAD/CAM potential would be essential to maintaining a competitive U.S. industrial posture and, that the technology would have greater impact on national economic well being than any other available technology. The academy's view was that 80 percent of design, planning, and shop operations plus 60 percent of assembly could be made substantially more efficient by applying CAD/CAM. Since 1977, numerous industries and organizations have made major commitments to CAD/CAM, such as GM and Ford, AT&T and ITT, NASA, large computer manufacturers, the aerospace industry, the Air Force and Army, private shipyards, and key industries in Japan. To now, the Navy has been notable by its absence from this group.

Experience by these and others since 1977 has confirmed the Academy's view of the urgency of the need for automation. As the quotations below suggest, the urgency often shifts to alarm about the challenges, and threats now confronting U.S. industry.

> "U.S. status as a first rank industrial power is at stake."
> (David Packard, Chairman, Hewlitt-Packard)

> "The choice that confronts American industry is to automate, emigrate, or evaporate."
> (James Baker, Executive Vice President, General Electric)

With somewhat more restraint, the National Research Council has advised that, "a joint shipbuilder-Navy (CAD/CAM) program should be set up..."

The Senate Appropriations Committee has requested a report on the potential expansion of computer aided design and manufacturing techniques at Naval shipyards and engineering centers. The Committee was aware that some private shipyards had reported 20 to 30 percent reductions of targeted costs. Yet the Navy investment in CAD and CAD/CAM was much less than in ADP systems - $200 and 800 million respectively, compared to $5 billion for ADP. CAD/CAM offers a much greater potential cost gain, at least 20%, while ADP systems yield less than 10%. The cause of this disparity in investment is that the cost for a major CAD/CAM system might range to several billion dollars, and that of a typical business ADP system to 10 million.

Private sector experience with CAD/CAM is compelling and supports the case for Navy automation. In the U.S. automobile industry for example, in 1978, there was computer-aided design, tooling, material handling, and some production. From 1980-84, automakers invested 60-80 billion dollars in CAD/CAM. When first quarter results of 1978 and 1984 were compared, production was up 2%, quality was up, and there were 23% fewer workers. 20 billion more has been invested in automated framing and testing at Buick City, where labor is down 38% from 1984 levels. In a broad spectrum of other industries, there are reports of substantial reductions in cost, time, labor, and failure rates.

From the Navy's review of private sector experience, four lessons have emerged, not all of them positive: (1) Installations of large CAD/CAM systems required strict standards and strict compliance with standards. (2) Data management system limitations stalled some installations (Boeing and GM programs, for example, where corrections cost $260 million plus rework). (3) There is the potential for very large payoffs from integration--from feeding design output directly to machines. (4) Adequate preparatory training of engineers is a critical, common factor necessary for success, but one often ignored.

Implications for the Navy of special interest from these preliminary studies were that (1) CAD/CAM operating costs could replace substantially larger costs associated with current methods--the essence of discussions within the Navy is cost reduction; and (2) the rate of cost reduction that might be achieved would be determined by the rate of investment in CAD/CAM. In the case of a projected 4:1 three-year payoff, which is likely with typical CAD/CAM gains, the ten-year net cost effect (return minus investment) would be plus 10 billion dollars when the investment rate is 100 million dollars per year, or 4 billion dollars when the investment rate is 40 million per year. These investment amounts correspond to sensible high and low limits for an initial CAD/CAM program.

The status of the NICADMM program is as follows: (1) Many of the administrative hurdles that a new Navy program must clear have been passed: liaison with Army ECAM and USAF ICAM programs has been established; a NAVMAT CAD/CAM liaison group has been operating one year; the Navy program has been included in the Department of the Navy Information Systems Plan of June 1984; a National Research Study on the Navy automation proposal is being published; a Program Management Plan has been completed, and an acquisition plan and appendixes are being prepared. (2) CAD equipment has been installed in 36 Naval activities, and the percentage installed is in the 70 to 90 percent complete range for all affected systems commands. A larger increment of installed equipment is being planned. Effects obtained from initial CAD/CAM installations in NAVSEA include decreases of 20-50 percent in labor costs, design manhours, two dimensional layout and drafting time, and bills of material. As significant as these gains are, the most important effect is that Navy engineers are gaining CAD/CAM experience. (3) A data exchange standard has been chosen and installed: Initial Graphics Exchange Specification, developed by the National Bureau of Standards as the *de facto* national standard, was invoked in 1984, and will acquire needed shipbuilding additions. It was invoked by GM internally and for all its suppliers in 1983, and it will become a standard for all Navy CAD/CAM.

During the first of its 'plan years', in FY 85 or FY 86, the Navy will undertake a considerable number of projects, particularly the establishment of standards for data management systems, operating systems, languages, graphic software, software tools, mainframes, CAD/CAM systems, professional computers, drafting equipment, and data exchange (already accomplished). In addition, central parts of the program will go forward, including analysis, design, programming, system testing, documentation, implementation, acceptance, evaluation, maintenance and alteration of equipment. During the 'second plan year', many of these separate projects will be combined, with project synthesis in material handling and control, drydock operations, cabling and wiring, hull structure, sheet metal operations, piping CAD/CAM, alterations CAD/CAP (computer-aided planning), machine shop operations, and so on. The essence of the second year's work is to select a combination of projects for the third to seventh years and to develop the plans for creating these increments. Evaluation of effects on quality, overhaul cost, time in yard, investment, payoff schedule, and preparation will be carried on simultaneously with this development.

Budget requests will increase accordingly, from $5 million in FY 85 for standards and criteria work, $14 million in FY 86 for synthesis and evaluation, $27 million in FY 87 for operations and management, and $58 million in FY 88. If results after FY 88 are not substantial, recommendations will be made to cancel the program. If they are substantial, recommendations will be made to reinvest some of the savings in order to accelerate net cost reduction. Outside the Navy, private shipyards will play a central role in this accomplishment of automation. For ship design, construction, and shipyard maintenance, potential cost savings are very high (highest for construction). Half of shipyard maintenance, most ship design, and all ship construction is carried out by private shipyards. Private shipyards personnel have three times as much CAD/CAM experience as Navy personnel, but private shipyards have realized only one third of the gain possible from CAD/CAM. Thus the Navy effort will involve improvement of the use of CAD/CAM in private shipyards. The first joint action will be to develop a data base design covering the life cycle of a ship class. The private shipyard contribution will be unique for each shipyard, but it is expected that 80 to 90 percent of the first design will be reusable in later programs. An additional topic of concern is whether the Navy should fund part of the shipyard's initial cost in anticipation of lower total costs.

Decisions are pending on all of these funding matters, but the Navy effort goes forward (slowly, because of the unprecedented levels of investment required). Minimum significant impact with an investment of $400 million may be a return of $1.6 billion. Maximum cost improvement may be 25%. Two or more large private shipyards will be involved at a cost for each of $200 million. There will be four or more years until the lead ship award. By that time, NICADDM should be in place, and a new era in Naval design, construction, and maintenance will have begun.

BIOGRAPHY

CAPTAIN JOHN F. LEAHY, III, USN

John F. Leahy, III, served as Program Manager Information Systems Improvement Program (PMS 309), Naval Sea Systems Command, Department of the Navy, Washington, D.C. He was also the Program Manager, Naval Material Command Computer-Aided Design and Computer-Aided Manufacturing (CAD/CAM) Program. He managed all computer-based systems used by the Naval Sea Systems Command. CAD/CAM efforts for that segment of the Navy responsible for design, acquisition, and maintenance of all Navy material used at sea or ashore.

Captain Leahy obtained a B.S. degree in engineering from the U.S. Naval Academy, Annapolis, Maryland, and an M.S. degree in computer science from the U.S. Naval Postgraduate School, Monterey, California. He has lectured extensively in computer science, mathematics, and business management at both the undergraduate and graduate levels at several universities, and is currently an adjunct professor in computer science and office automation for Virginia Polytechnical Institute and State University. Captain Leahy is a member of the American Society of Naval Engineers, the Association of Information System Professionals and the National Computer Graphics Association. He is also a member of the Advisory Board for Defense Computers-Graphics.

He served in the Submarine Force, U.S. Atlantic Fleet on the USS Becuna and the ballistic missile submarine USS George Washington and as the Submarine Force Plans Officer. Later he served at Puget Sound Naval Shipyard as the waterfront production superintendent for the conversion of a ballistic missile submarine and as the planning and repair superintendent for modernization and refueling of a nuclear-powered cruiser. Then he managed all industrial functions in the submarine tender USS Proteus. In his last assignment before being assigned to Washington, he was the Deputy Supervisor of Shipbuilding, and Trident Project Officer at Groton, Connecticut. In this position, he manged the acquisition of 34 warships. In Washington, he served in corporate planning and as the executive assistant to the Commander, Naval Sea Systems Command before being assigned to his present position.

15

PERSONAL WORKSTATIONS FOR CAD/CAM APPLICATIONS

Robert W. Lingard, Ph.D.
CADAM INC
1935 N. Buena Vista Street
Burbank, California 91505 U.S.A.

Abstract

For CAD/CAM applications, the personal workstation offers significant advantages over the time-shared, central computing facility accessed through a graphics terminal. Among these advantages are availability, consistency of response, portability, and the ability to dedicate significant resources to a single task without affecting other users. Recent advances in hardware technology have made the powerful personal workstation not only a practical reality, but an economically advantageous option for CAD/CAM tasks.

At CADAM INC, we are moving many CAD/CAM applications to the personal workstation environment. Our goal is to provide users with a complete environment in which highly interactive CAD/CAM tasks can be performed on a workstation. That same workstation can be connected to a large central computer for database management or complex computational tasks. The end result will be an environment that enhances individual user effectiveness and improves the overall productivity of an organization.

1. INTRODUCTION

The key to the success of today's CAD/CAM software systems lies in their ability to greatly increase the productivity of their users. By using the computer to automate many of the design, analysis, drafting, and manufacturing tasks previously done by hand, CAD/CAM systems have had a profound impact in reducing overall production costs in many industries. The current progress in computer technology, with the resulting decrease in computing costs, will have an even greater effect in increasing productivity.

The most significant phenomenon made possible by modern semiconductor technology is personal computing. Decreasing hardware costs have made it economically feasible to dedicate significant computer resources to a single user. Although there is no universally accepted definition of personal computing, the description proposed by Raskin and Whitney will be used here.(1) They describe personal computing as computing where, "the resource is personally . . . managed by an individual." This contrasts with time-shared computing where computer resources are simultaneously shared by several users and managed by a central computing organization.

Time-sharing, a concept introduced with second generation computers, dramatically changed computing by giving many individuals interactive access to large powerful machines. The reason for sharing the resource was simple--cost. At millions of dollars each, these computers had to be shared. Now that significant computing power and memory can be provided relatively inexpensively, time-sharing is no longer required for many computing applications. The capital investment that today provides each user with a small slice of a time-shared machine can provide that same user with his own powerful machine equipped with a high-resolution, color graphics display. Newell, Fahlman, and Sproull go so far as to say, "The era of the time-shared computer . . . is drawing to a close . . . . We are entering a new world of powerful personal computers, as different from the world of time-sharing as that world was from batch processing."(2)

2. THE PERSONAL WORKSTATION

The concept of personal computing as it has been defined here is independent of the size of the computer used. It makes economic sense, of course, to dedicate only relatively inexpensive computers to individual users. Still, there is a wide range of computers that could be used cost effectively as personal machines. They range in price from thousands to tens of thousands of dollars. For most CAD/CAM applications, computers at the higher end of the scale are usually required. The term "personal workstation" is used here for such a computer. This term is used to make a distinction between personal computers as they are generally envisioned and the more powerful personal machines required for CAD/CAM. A typical personal workstation would likely have the following features:

- a speed of one million instructions per second
- a main memory size of two megabytes
- about 50 to 100 megabytes of local secondary storage
- a high quality (1000 by 1000 pixel) color display device
- a pointing device (light pen, tablet and stylus, mouse, etc.)

- a virtual memory multi-tasking operating system (possibly UNIX)
- high level programming language support (C, Pascal, FORTRAN, etc.)
- a high-level graphics interface (probably based on the Graphical Kernel System)
- a local area network connection
- possibly a locally attached hardcopy device (printer or plotter)

Although not strictly required by the definition of personal computing as put forth here, the personal workstation, as we view it at CADAM INC, has two important additional features. First, it is a general purpose machine. It is capable of running applications other than CAD/CAM, and it can support end user program development activities. This concept is in conflict with a turnkey approach to CAD/CAM where a complete ready-to-use hardware and software package is provided by a single vendor. It is not generally expected that a user will add other capabilities to his turnkey package; and capabilities for user programming are not usually provided. The advantages of the turnkey approach are expressed by Orr; he states that the turnkey approach ". . . is often the easiest way to implement CAD/CAM".(3)

Our reason for advocating a general purpose machine rests in the belief that it is impossible for a single vendor to provide all the software needed for CAD/CAM and CAD/CAM-related activities. Increasingly, customers are demanding that vendors develop interfaces between various CAD/CAM packages and even between CAD/CAM and other software systems. They want to be able to use the drafting system from one vendor with the solid modeling package from another or to use drawings created with their CAD/CAM tools within a text processing system. It is becoming increasingly difficult for a single vendor to provide all the software that a user requires on a workstation. Providing a general purpose workstation allows the customer to acquire or develop the other needed software and software interfaces between application packages.

The second feature that we have imposed on the personal workstation is that it must be usable within a normal office environment. Ideally, it should be able to run within a typical office environment in terms of its electrical and air conditioning requirements; and it should be quiet and small enough to fit on, or next to, a user's desk. For the workstation to be a truly personal resource, the user should have easy access to it when it is needed; this means it must be located where most other work is accomplished.

In summary, we have defined the personal workstation as a powerful, standalone, general purpose computer that operates as a dedicated resource for a single individual and is located in the user's personal work space.

3. ADVANTAGES OF PERSONAL COMPUTING

A personal workstation, as described above, offers significant advantages over the time-shared approach for CAD/CAM applications.

RESPONSIVENESS

The first and most important advantage of a personal workstation is responsiveness to the user. The importance of good response cannot be stressed too much. Many studies have been conducted on the effect of system response on programmer productivity. It seems safe to assume that those results can be applied to CAD/CAM users as well. In a recently published study by Lambert of IBM, it was shown that reducing the average system response time from 2.22 seconds to 0.84 seconds increased the productivity of software developers by 62 percent.(4) While these numbers may not be directly applicable to CAD/CAM system users, it is clear that response time does have an effect on productivity. Studies conducted at CADAM INC have shown that user productivity on CADAM® continually increases with decreases in response time until trivial operations are performed in less than 0.5 seconds. Trivial operations are those that add or modify geometric elements (e.g., draw a circle or trim a line). Some customers claim that they have further increased productivity by reducing the response time for trivial operations to 0.25 seconds.

Maintaining good response in a time-sharing environment is a challenging task. To achieve acceptable response during peak periods usually means providing significant unused resources during off hours. Most companies find it difficult to justify such an expense. The result for the user is inconsistent response. In the middle of the night, response is probably excellent, while during the day it is likely to be poor. During even a single session a user might experience widely varying response times as the load on the system changes. This inconsistency of response has been shown to be more harmful to user productivity than consistently poor response.

The personal work station essentially eliminates these response problems. The level of response desired can usually be achieved by purchasing the appropriately powered hardware, and consistency is guaranteed since there are never competing tasks on the system.

RELIABILITY

The second advantage of a personal workstation environment is its inherent reliability. A failure in any one machine cannot bring down the others. The critical demonstration or production work can simply be moved to a working machine. The only critical resources are the shared secondary storage and hardcopy output devices. Generally, backups can be provided for files on secondary storage and alternate output devices can be made available. Therefore, one malfunctioning computer will not usually impact critical tasks.

EXTENDABLITY

A third advantage is extendability. The personal workstation environment can be extended in much smaller increments than a comparable time-shared system. A few new users can be easily accommodated by adding a few new workstations. An installation will not have to make difficult decisions between running with one overloaded machine or two underutilized ones. An important corollary to this is that a small installation with only a few

® CADAM is a registered trademark of CADAM INC.

users can provide essentially the same environment for these people as larger companies provide for hundreds of users. This should make it possible for CAD/CAM software to be used productively in many places that today cannot reach the critical mass of work necessary to justify the purchase of a large computer.

PORTABILITY

Portability of the workstation itself is yet another advantage. Most high-resolution graphics scopes, which are attached to today's large computers for CAD/CAM applications, are very difficult to move from one location to another. There are usually severe distance restrictions that make it difficult for the display device to be placed very far from the host computer. Expensive communication links must be provided, or response time suffers greatly. Unless conventional telephone communications are used, moving a display device to a location far from the host computer requires significant effort, time, and money. With the personal workstation, however, moving a CAD/CAM capability to a new location can be easily accomplished by simply moving the entire machine. This gives companies greater flexibility for providing CAD/CAM capabilities where required as workloads and the needs of various projects change over time.

SECURITY

When users are working at terminals connected to large time-shared computers, providing an environment with the necessary security protection for military or other sensitive work can be difficult. Either the entire computing facility must be secured or communication lines must be encrypted. Such solutions can be costly and difficult to implement. With the personal workstation, security problems can often be solved simply. Since the personal workstation is a stand-alone computer, it is easy to provide a user with a secure computing capability by merely placing the workstation and the user within a facility that is secure.

SIMPLICITY

Since the personal workstation is a single-user system, the software written for it can be simpler. Sophisticated CAD/CAM systems like CADAM, in order to operate in time-sharing environments, must contain significant amounts of code to manage shared data and to protect users from one another. On a single-user, system this programming effort is not required. The result is a software system that is easier to develop, easier to maintain, and one which operates more efficiently.

CONTROLLABILITY

A significant advantage of the personal workstation is that each user gets the machine's full resources regardless of what other users are doing. If a user chooses to run a big job, he can do so without impacting other users. This is especially important when more sophisticated and resource-demanding CAD/CAM applications are used. Solid modeling and automated numerical control are examples of such applications. In time-shared environments, one user performing such a resource-intensive task could affect the response times of all other users on the system. On the personal workstation, only the user performing the task is affected. He has complete control over his computing resource and can make decisions freely about its use without worrying about the effect on other users.

COST

The final advantage of the personal workstation is cost. The combined average hardware and software cost for a typical CAD/CAM workstation is in excess of $100,000. A personal workstation with equivalent capability can be provided for about one half that cost, and the cost of workstation hardware is continuing to drop. In addition, personal workstations are easier to maintain than large machines, and on-going operational costs are lower.

Since the personal workstation is a complete computing system, a company can start with a single machine. There is no need to wait until several workstations can be justified before purchasing a system. This broadens the market for CAD/CAM systems tremendously and can spread the productivity gains from such systems to smaller and smaller companies.

4. CHALLENGES FOR THE PERSONAL COMPUTING APPROACH

The advantages of this approach are clear. It is not, however, without some problems. Four major difficulties, outlined below, are communications, database management, organizational control, and overall system complexity. These problems, either created by or amplified by personal computing, must be dealt with if the full potential of this approach for CAD/CAM applications is to be reached.

COMMUNICATIONS

The major problem that personal computing aggravates is communications. Providing each user with his or her own self-contained CAD/CAM computing capability is fine, but few users work in total isolation. Generally, there is a need for one user to obtain information from and provide information to other users. If users are close enough together and use the same type of hardware, a good local area network can solve the problem. If, however, distances are great or users are on incompatible systems, the transfer of information can be difficult to accomplish. This problem, of course, exists with or without personal workstations, but an increase in the number of independent computers within a company can have a tremendous impact on communications needs.

Presently, there is a lack of both the necessary hardware and software to adequately address this situation. The level of communication that can be achieved today between dissimilar systems is primitive. Inexpensive solutions are slow and awkward. Even expensive, high-speed solutions are difficult to implement and are not generally applicable. Although significant research and development is underway in both hardware and software for the communication of computer-generated data, a general solution cannot be expected for some time. Successful solutions today are ad hoc. Usually a company tries to minimize its problems by buying hardware only from a single vendor or develops specific interfaces between the different pieces of equipment it has in use. The lack of convenient, fast ways to exchange information between different systems over long distances will probably be the major factor inhibiting large companies from adopting a personal workstation approach for CAD/CAM applications.

DATABASE MANAGEMENT

Transferring data between computers is only part of the problem. Keeping track of all the data for a project or for a company becomes exceedingly difficult when it is distributed among multiple machines. Effective means for sharing and managing data on personal workstations connected by a local area network have been developed, but when data is spread over dissimilar machines or great distances, management of the information is a problem.

Although most CAD/CAM systems today provide file management, or even true database management capabilities, they generally operate only in single processor environments. One solution is to have one large computer that operates as a database machine. If it is required that all other machines checkout and return data to that computer, an organization can achieve central control of its data. This requires that all personal workstations be connected either directly or indirectly to the database machine and that procedures be established or software developed to insure the integrity of data in the system. For example, if one user on one workstation obtains some data for modification from the database machine, somehow the system must prevent other users on other workstations from getting that same data until it is returned by the first user. Few systems today provide this level of control. It is also clear that managing all data on a single computer may not be practical because of complexities, inefficiencies, and costs associated with the implied communications requirements. Solutions for managing data across multiple machines must be developed. The effective management of a distributed database is one of the most important challenges for the future.

ORGANIZATIONAL CONTROL

Related to the issue of database management, but more general in nature, is the problem of organizational control. In order to provide an effective and efficient computing environment for an organization as a whole, including the protection of important company information, a certain amount of central control is necessary. If each user organization can independently acquire its own personal workstations, managing and controlling the company's CAD/CAM data can be difficult. Without an in-depth examination of the issues of centralized versus decentralized computing, suffice it to say that the personal workstation approach makes any kind of centralized control difficult.

A comprehensive examination of the advantages and disadvantages of decentralized computing was conducted by King.(5)

OVERALL SYSTEM COMPLEXITY

Providing general purpose personal workstations capable of communicating with a variety of other machines creates a complex environment for users as well as for the organization charged with maintaining the overall computing environment. Although this environment provides users with great flexibility, it is often not easy to take advantage of that flexibility. Access to a wide variety of software is appealing, but few interfaces exist between systems. Furthermore, the user interface to each CAD/CAM package is different from the others. Finally, from the systems maintenance point of view, having to deal with many different hardware and software vendors adds significant complexity to the task of supporting the user community.

Although the effort in providing this additional flexibility is probably justified, there is much room for improvement. If standards for representing CAD/CAM data could be established and used, the transfer of information from one system to another could be greatly simplified. The Initial Graphics Exchange Specification (6) is an example of such an effort. Although incomplete and not universally accepted, it is an important step in this direction. If similar efforts were to address user interfaces, significant progress could be made toward allowing users easier access to various software systems through a single personal workstation.

5. THE APPROACH AT CADAM INC

At CADAM INC, we have been working to provide a version of the CADAM System on a personal workstation. Initially only a subset of CADAM is being developed. However, based on what can be expected in workstation computing power, it should be feasible in the future to provide full CADAM capability in that environment. CADAM on a personal workstation, or Personal CADAM as we have named it, is being developed as one element in a complete CAD/CAM environment. It will enable installations to easily provide CADAM capabilities at remote locations or in secure environments and will provide a less costly solution where only one or two workstations are required. Personal CADAM is being planned as a system which can function independently or as one capability among several CAD/CAM systems. The ability to exchange information with other software systems will be provided; and specific interfaces between Personal CADAM and other CAD/CAM packages are being planned.

Data management for large installations will continue to be the function of the large central computer. Personal CADAM will have extensive capabilities for the exchange of information with a central computer in addition to capabilities for workstation-to-workstation communication through a local area network. As a fully compatible element in a CADAM environment, Personal CADAM can help to overcome some of the shortcomings of the time-shared approach for CAD/CAM.

6. CONCLUSION

There is no doubt that CAD/CAM systems have had a significant affect in improving productivity in many engineering and manufacturing activities. The personal workstation offers the potential for even greater productivity gains by making it cost effective to extend CAD/CAM capabilities to far greater numbers of users. Furthermore, because of the unique characteristics of the personal workstation in terms of responsiveness, availability, and control, each CAD/CAM system user can be more productive. Although some problems relating to communications, database management, and interfaces between systems need urgent attention, it is clear that the personal workstation will play a significant role in CAD/CAM systems of the future.

7. REFERENCES

1. Raskin, J. and Whitney, T., January 1981, Perspectives on Personal Computing, IEEE Computer.

2. Newell, A., Fahlman, S., and Sproull, R., July 1979, A Proposal for Personal Scientific Computing, Department of Computer Science, Carnegie-Mellon University.

3. Orr, J., December 1984, Turnkey Systems Provide Single-Source Solutions, CAE Computer Aided Engineering, Penton/IPC, Inc.

4. Lambert, G., 1984, A Comparative Study of System Response Time on Program Developer Productivity, IBM Systems Journal, Vol. 23, No. 1.

5. King, J., December 1983, Centralized versus Decentralized Computing: Organizational Considerations and Management Options, ACM Computing Surveys, Vol. 15, No. 4.

6. Smith, B., et al, 1982, Initial Graphics Exchange Specification (IGES), Version 2.0, NBSIR 82-2631 (AF), National Bureau of Standards.

8. BIOGRAPHY

Dr. Lingard is currently director of the Personal CADAM Business Unit of CADAM INC. CADAM INC is a wholly owned subsidiary of the Lockheed Corporation. He joined the company in 1981, and assumed the position of director of planning and quality assurance. Previously, he was on the staff of the Information Sciences Institute of the University of Southern California and was involved in artificial intelligence research.

Dr. Lingard has authored several papers on software engineering and artificial intelligence. He has taught at both California State University, Northridge, and the University of Southern California where he received his Ph.D. in Computer Science in 1975.

16

COMPUTER GRAPHICS FROM THE GRAPHIC DESIGN PERSPECTIVE

Aaron Marcus, Principal
Aaron Marcus and Associates
1196 Euclid Avenue
Berkeley, California 94708 USA

1. INTRODUCTION

Programmers and managers of all kinds of workstations in CAD/CAM, engineering, office automation, computer-aided instruction, and programming environments can benefit from the principles of effective, systematic graphic design (visible language programming) applied to output formats and to screen interfaces. The following paper summarizes some important points that have emerged from three years of active involvement in designing computer graphics for a variety of systems and many different kinds of users. Although the discussion emphasizes high-resolution, dynamic interface screen design issues, the principles in general apply to static laser printer, dot-matrix, and plotter output as well. The final section of the paper outlines a future development scenario for system development firms in order to acquire expertise in visible language programming.

2. TYPOGRAPHY

Typography is one of the most obvious areas in which graphic design can asist computer graphics display design. A better use of available type fonts can promote legibility and readability. Users are constantly searching for or creating information. Concise, consistent typographic cues to the current location of the user, to the current structure of information, and to the location of other components of data or functions are crucial to users' being effective and at ease when working with computer-generated displays.

The typography needs to be usefully redundant. In menus, lists, online help, or other documents, changes of typefont, type slant (roman or italic), boldness, and location can aid users looking at a complex display. The largest lettering can be as much as five times the normal text type size for selected items. The smallest lettering should be large enough for easy viewing in available light and at normal viewing distances for screen or paper display.

All software modules should be part of a systematic approach to screen interface design for a workstation. This approach also applies to all document design for a workstation. Once users have some experience with part of the system, its basic consistency will allow them to become productive more quickly when using other parts of the software or when checking new components of documentation.

Concerning font design, better type faces are required for most systems. There is a particular need for new typefaces to enhance the readability of special texts, such as source code. There is also a need for improved legibility when text appears against a black background or a gray background, as is often the case for screen or page highlighting of text components such as titles or commands. Certain pixel patterns tend to make some small fonts hard to read in black and gray areas unless they are designed properly. The appearance of these fonts on screens and laser printers is very important because screens and paper documents can become more readable with black or gray areas to guide the viewer's attention. Another factor of typographic importance is the degree of difference in bold type and italics from standard medium roman fonts. The weights and differences of letterform must be strong enough to be clearly distinguished on screens and printers that typically may be poorly adjusted.

3. SYMBOLISM

The various means of high-lighting, low-lighting, and otherwise distinguishing areas of a screen or page contribute to the symbol system of a workstation and its documentation. In addition, small individual symbols, some of which are sometimes called icons, can efficiently communicate messages that would otherwise require many alphanumeric characters to display.

For screens, all icons, cursors, and window titles should be carefully studied so that a minimal set of clearly distinguished symbols can be designed to communicate all the necessary functions, states, and other annotations. Much of this information can be communicated in a pictographic manner that facilitates recognition and memorization as well as providing a source of user interest and attention. For example, cursors can change for different modes of the system. By using such changes, novice users can be helped to learn the system more easily and swiftly.

It is important for training purposes and for online help to design diagrams of the system to aid users in understanding the whole system and its parts. In general, any workstations and their documentation make too little use of this verbi-visual approach to communicating complex structure and process.

4. COLOR

One of the most significant color aspects is the controlled use of value (gray level) in depicting three levels of information in a screen or page: high-lighting, neutral state, and low-lighting. Because color fidelity is often poor (screens and printers may not be adjusted perfectly), the gray value differences will be one of the strongest and most durable aspects of differentiation. For black-and-white display environments, of course, this is the only dimension of color differentiation.

At the present time, multi-color, multi-window environments often show too many different colors with too-strong saturation or chroma. Such displays can cause confusion in the eyes and in the minds of users. More subdued, systematic approaches to color are required.

5. LAYOUT

Screens and pages can quickly become crowded and confusing to read through, even if there is relatively little on the screen. Systematic, effective design suggests a minimum of differences in column widths and illustration or chart widths. A layout grid for all typical components of the screen or page must be prepared. This will aid system or application programmers to layout material in a systematic manner. It will also help the user by subtly organizing all visual elements that must be examined.

One of the typical problems of screen and page layout is that there is not a stong visual hierarchy of elements. When users look at a screen or page, it should be visually obvious where one is to look first and where the most imporant parts of the display are located. Often many of the text elements in the screen or on the page seem to "float." This can be improved with systematic placement according to a grid which "locks" each element into its regular position. Module, chapter, frame, and page numbers of screen or documents need to be clearly labeled and should be shown visually in a strong manner.

Care must be taken when designing the layouts that the parts fit together well. For example, menu option selections or text columns should not be so close that they run together. When options on menus are selected, the inside area as well as the outline should reverse white to black; otherwise the reversed interior portion of the menu button looks smaller than its companions.

Lists are especially difficult to lay out well. Too much space often appears between the elements that should be grouped tightly together for quick and accurate reading. Ruled lines and gray areas can help, but these must be used carefully; otherwise they tend to make the lists in menus, windows, or pages look too busy and disorganized.

In general, there is a need for templates to help the system or application programmer (and eventually users) to quickly create the layouts that are used very often. In fact, there is a need to study what generic screens and documents should be supplied with the system. It seems advisable that the size of documents on the screen should be the same size as on the laser printer. On many systems currently, this is not the case. Many American high resolution display systems (Apple, Xerox, Apollo, Sun, Perq) offer horizontal format screens. This provides an adequate space for a depiction of typical vertical page on the screen plus the necessary menu items or window headers that most workstations require for document preparation, editing, and review.

6. SEQUENCING

Sequencing is one of the most subtle aspects of human-computer communication. Many of the best film directors know that the pacing and juxtaposition of imagery is a key ingredient of communicating facts, concepts, and even moods. In screen design, the display usually changes quickly. There may be times when slow speed is useful, especially for novice users. All items that are re-painted or re-displayed should appear in a logical manner, so that users are constantly reminded of the inherent structure of the display. For example, when the command names and their outline rectangles change, the name and the rectangle should change as a logical group.

Complete redisplay of the high-resolution image may require several seconds. Even a 5 second redisplay time may seem painfully long. If the screen image is structured suitably, then the parts that are visible can be examined and used while the rest of the image is filling in. This may make delays in display more acceptable.

7. INTERACTION

Once optical scanners and character recognition capabilities are added to system, a large amount of keyboarding can be removed from users' tasks. The editing and reviewing of documents, especially if they are generically structured, can be handled efficiently by mouse, icon menu, and visual editing actions. This will be an important change of interaciton paradigms, especially for work environments in which a large quantity of documents already exist that have not been keyboarded.

When editing images, graphics, and text, there is a need for "rulers", temporary grids, or other positioning tools to aid in alignment, determining proportions, or controlling the size of visual elements. These are basic and necessary rools that help make any kind of document editing workstation efficient and useful.

In archiving activities, it is possible to store many pictures. If there is no way to put comments with file names and file size, it may be difficult for users to know what is in the file. Users should receive online or offline help in storing files of graphics and images so that the file list is created in a logical and useful manner. It is also important the users be able to search a large collection of files by subject or, someday in the future, by visual attributes such as color, shapes, symbols, etc. It should be possible to do multiple kinds of searches of directories and files, for example, with attributes of text, picture elements, dates, and/or authors. This will be necessary especially when the collections of titles become large. One possibility is that the workstation could make small "icons" or abstract miniatures of the documents. In this way users could recognize the document by its general appearance even if the image on the screen is illegible.

There seems to be a need for global interaction commands that would make text editing, image editing, and data management functions seem less confusing, especially when users have access to more than one system. At some point in time generic functions and their generic visualization need to be established.

One point to bear in mind during development of effective systems is that it is very important to allow developers to make screen dumps so they can study interaction when building the system. Very often software and hardware of unfinished systems (and some finished systems) do not permit easy screen dumps. These are essential for "visually debugging" screen apearance, layout, and interaction from the graphic design or visual communications point of view.

There are other graphic approaches to the study of interaction; for example, a highspeed videotape can be made of a user at work to observe the patterns of keyboard and mouse use. Another important practice is to construct storyboards out of theoretical or working screens. In this way the developers can get a feeling for the entire system: its features, its inconsistencies, and its overall impact on the user.

8. CONCLUSIONS AND FUTURE SCENARIO

Based upon the graphic design work, or visible language programming, that Aaron Marcus and Associates has undertaken over several years, the concepts and advice mentioned above can assist system builders in understanding the issues involved in achieving better human-computer communication.

AM+A recommend that firms must prepare their own staffs for visible language programming activity. In the interim, a team of people with human factors, industrial design, and programming experience could accomplish this activity. Eventually, a graphic designer with specialized experitise in visible language programming should be hired and trained once the concepts and methodology of the team are established and resource literature has been assembled.

One of the first products that such a group can prepare is a corporate graphics design manual that clearly specifies and shows typography, symbolism, color (gray value), layout, animation, sequencing (pacing), sound, and interaction for all screens and functions. The document will enable many different engineers and programmers to build consistent modules, will facilitiate communication between engineering and marketing, and will improve the user's enjoyment and effective use of the workstation.

This paper described the potential and future development for visible language programming in the workstation development cycle. The accompanying references outline issues, method, and content in greater detail.

9. REFERENCES

1. Marcus, Aaron, "Designing the Face of an Interface," IEEE Computer Graphics and Application, Vol. 2, No. 1, January 1982, pp.23-29.

2. Marcus, Aaron, "Graphic Design for Computer Graphics," IEEE Computer Graphics and Applications, Vol. 3, No. 4, July 1983, pp.63-68.

3. Marcus, Aaron, "Corporate Identity for Iconic Interfaces," Computer Graphics and Applications, Vol. 4, No. 11, December 1984, (in press).

4. Marcus, Aaron, Michael Arent, and Burce Browne, "A Visible Language Program for Perq's Accent Operations System," Proc. National Computer Graphics Association Annual Conference, April 1985 (in press).

10. BIOGRAPHY

Aaron Marcus is the principal designer of Aaron Marcus and Associates, an information-oriented graphic design firm, specializing in effective communication for computer graphics display. Mr. Marcus has written and lectured extensively on these subjects. He co-authored The Computer Image, published by Addison Wesley, and wrote Managing Facts and Concepts, published by the Design Arts Program, National Endowment for the Arts. He is on the advisory boards of Computer Graphics Today and Information

Design Journal. His firm has consulted internationally with major computer graphics research and development groups on the subjects of chart, diagram, and map design on user-friendly iconic interfaces, and on program visualization/documentation techniques. Mr. Marcus is a member of ATI, Siggraph, NCGA, ACM, and the American Institute of Graphic Arts.

17

THE DIGITAL MAPPING PROGRAM OF THE U.S. GEOLOGICAL SURVEY

Alan M. Mikuni
U.S. Geological Survey, National Mapping Division
Western Mapping Center
345 Middlefield Road
Menlo Park, California 94025, U.S.A.

ABSTRACT

As the nation's primary Federal earth-science, factfinding and research agency, the U.S. Geological Survey (USGS) provides cartographic and natural resource information to many users to aid in planning and decision-making. Users of the data, both inside and outside the USGS, have become more and more sophisticated in their applications. As a consequence, scientists at the USGS are embracing computer technology as an essential element of their data gathering and production activities. Now, much of that cartographic information is available in computer-compatible form on magnetic tape from USGS national digital cartographic data bases.

INTRODUCTION

Before discussing the USGS' computer data bases and automated cartography and how they relate to the management of natural resource data, it may be useful to briefly review the mission of the agency.

The USGS was established in 1879 and was made responsible for the "classification of public lands and examination of the geological structure,

---

Publication authorized by the Director, U.S. Geological Survey, on April 30, 1985.

Note: Any use of trade names or trademarks in this paper is for descriptive purposes only and does not constitute endorsement by the U.S. Geological Survey.

mineral resources, and products of the national domain." Since that original charge, a number of changes, including increased awareness of man's impact on the environment, the need to carefully manage the use of nonrenewable resources, and the search for new sources of critically needed energy and mineral commodities, have caused new responsibilities to be added to the mission of the Geological Survey. Among these are topographic and geologic mapping, geographic research, hazards monitoring, and water resources assessments.

The following comments will be directed toward describing the digital cartographic data bases established and maintained by the National Mapping Division (NMD) of the USGS, and some of the methodologies employed by the NMD in the development of those data bases.

NATIONAL DIGITAL CARTOGRAPHIC DATA BASE

In response to the expressed requirements of major users, especially other Federal agencies, the NMD is producing a number of digital cartographic products. Digital elevation data, land use and land cover data, and planimetric data are being developed and placed in a repository known as the National Digital Cartographic Data Base (NDCDB). Collectively, these data have been given the name "US GeoData" by the USGS to provide an overall identity to the variety of products being developed.

Two major classes of digital data presently available reflect the capabilities of the USGS for data production. They are the planimetric data, digital line graphs (DLG), and the elevation data, digital elevation models (DEM). Three types of DLGs are presently available: (1) the 1:24,000-scale topographic series, (2) the 1:250,000-scale land use and land cover series, and (3) the 1:2,000,000-scale planimetric information series. A fourth type of DLG, the 1:100,000-scale series, is being developed as part of a cooperative project with the U.S. Bureau of the Census. DEMs are available as (1) 1:24,000-scale DEM data, and (2) 1:250,000-scale Defense Mapping Agency digital terrain data. More detailed information regarding these digital data, such as coverage available, costs, and user guides, can be obtained from the National Cartographic Information Center, headquartered in Reston, VA.

DIGITAL LINE GRAPHS (1:24,000)

Over 7 years of effort have been invested by the NMD in the development of techniques, procedures, and software for the capture and processing of map data in the form of DLG information from the 7.5-minute topographic map series. Basically, a DLG is line map information in computer-compatible form and includes planimetric base categories such as boundaries, transportation, and hydrography. In addition, a non-base category, the U.S. Public Land Survey System (USPLSS), is being captured in response to the requirements of major Federal data users such as the Bureau of Land Management.

Digitization of the graph elements known as nodes, areas, and lines from annotated base maps provides the raw data for the construction of a DLG. A node is defined as the end point of a line, and one or more lines can end at this single node. An area is a closed polygon, and is represented in the DLG realm by a single digitized point, generally, but not always, located within the polygon it represents. A line is a directed string of coordinate values extending between two nodes with defined areas to the left and to the right. Single point map features, such as wells, springs, or location monuments, are defined as "degenerate lines" or extremely short lines whose starting and ending nodes are within tolerance level proximity.

Each of these nodes, areas, or lines may have one or more attributes, or numerical descriptors, attached to it in the form of a list. Attributes describe the physical or cultural characteristics of that graph element within the context of the topographic map. In general, every cartographic symbol will have at least one unique code value. The tagging of a feature with multiple codes can occur, as well. The NMD uses a seven-digit attribute coding scheme composed of a three-digit major code and a four-digit minor code. The major code defines the map category (or layer) portrayed, that is, USPLSS, coded 300, and the minor code defines a feature within that overlay, that is, a section corner, designated 0001. Together they comprise the attribute code "300.0001" for a node representing a section corner. Manual X-Y digitizing systems, such as the Altek AC90SM, and interactive digitizing/editing systems like the Intergraph IGDS, are used extensively to capture raw data used to develop DLG files.

A DLG may be classified at one of three levels depending largely on the extent of editing, enhancements, or spatial structuring it received during processing.

DLG-1. DLG level 1 is the simplest structure that maintains the original (raw) data in a standardized format, coded to prescribed standards, and edited for normal data acquisition blunders. This level of information is designed to meet three general needs: (1) providing a quick source of digital data, (2) making raw data available to users who are capable of completing the structuring themselves, and (3) providing data for display or plotting systems of low or moderate cartographic quality.

DLG-2. DLG level 2 structure is designed to support graphic display and plotting systems of high cartographic quality. The data contain additional attribute codes and visible errors and inconsistencies are removed.

DLG-3. DLG level 3 is essentially DLG-2 information that has been spatially structured to define all topological relationships. Fully topologically structured data files like DLG-3 are designed to be fully integrated into geographic information systems.

Virtually all DLG data collected at the USGS is processed to level 3 using a software system known as UCLGES (Unified Cartographic Line Graph Encoding System). Data collected under the specifications and criteria of levels 1 and 2 cannot always be enhanced to level 3. But, the reverse process of extracting a level 2 DLG from level 3 or level 1 from level 2 data is more easily achieved. The additional cost of going through the exercise of topological and attribute coding to achieve level 3 DLG data pays handsome dividends to users, because they are provided with a highly versatile, yet relatively inexpensive, data file that will satisfy virtually all their requirements. The capability exists within a DLG-3 to support activities ranging from Geographic Information System-level applications to simple cartographic plotting.

Topology, as defined for digital cartography and spatial analysis, requires that all lines be explicitly joined to nodes and that all areas be explicitly linked to bounding lines (arcs). The topological structuring, or the spatial relationships of map features on a topographic quadrangle, are very specific when the graphic map is observed visually. These relationships become slightly less straightforward, however, when only the digitized coordinate values are known and the topology is NOT defined. Digital data without spatial structuring are nothing more than "spaghetti." The relationships of these coordinate strings to one another are defined only by the coordinates themselves. A graphic produced from the topographic "spaghetti" will obviously and promptly reinstate those relationships, but the overall utility of this type of digital data is limited to just that: graphics. A DLG-3 data file contains within the syntax of the coding (coordinate values and attribute codes) the desired topological structuring. An appropriately written software package will enable a user to perform spatial analyses and modeling without the need for a graphic map, simply because all relationships are explicit within the file. This, then, is the intent of the very highly structured nature of a DLG-3: to completely define all the graphic and topological relationships of nodes, areas, points, and lines on a map so that the resulting data file will prove useful to the most applications in the largest number of disciplines.

To date, over 20,000 DLG-3 overlays are resident in the NDCDB to help meet the combined needs of spatial data users.

LAND USE/LAND COVER (1:250,000-POLYGON)

Land use and land cover information in digital form is also available. "Land use" refers to the activities of man that are directly related to the land. "Land cover" describes the vegetation, natural surface, water, and artificial constructions on the land surface. This information is compiled from aerial photography and other remotely sensed data sources onto 1:100,000-scale and 1:250,000-scale maps, collected, and incorporated into the Geographic Information Retrieval and Analysis System (GIRAS). The source maps for the GIRAS activity are classified as polygon or areal maps. Each single-theme land use/land cover (LU/LC) map consists of non-overlapping areas (polygons) delimited by lines. In the context of LU/LC mapping, the lines have no meaning other than to delimit the areas, each of which is identified or classified according to a standard classification scheme. Once classified, the graphic map data on the LU/LC map are transformed to an optimal form using GIRAS.

The GIRAS format is topologically structured and is designed to process large quantities of polygon-type map data. The topological elements associated with polygon maps are nodes, arcs, polygons, islands, complex islands, and polygon labels. A polygon is an area that is homogeneous in the characteristic being mapped. An arc delineates the boundary either between two polygons, or between a polygon and the outside of the map. Nodes define the ends of an arc. A simple island polygon is completely surrounded by another polygon, and complex island polygons consist of polygons that share common arcs, also completely surrounded. A string of X-Y coordinates from a starting node to an ending node defines an arc. The sequence of the X-Y coordinates of an arc indicates a direction of "travel" which, since polygons are delimited by arcs, defines a polygon to the left and to the right. A polygon, then, can be spatially defined as the sequence of arcs that comprise its boundary, both external perimeter and any islands that lie within the perimeter. As with DLG structuring, the basic topological elements of a GIRAS polygon file are uniquely identified and cross-referenced to each other.

The LU/LC classification system includes codes to describe polygons that are:

- Urban or built-up land
- Agricultural land
- Rangeland
- Forest land
- Water
- Wetland
- Barren land
- Tundra
- Perennial snow or ice

These nine general Level I categories are further subdivided into 37 more specifically defined Level II categories. The minimum size for the mapping of areas is 10 acres for water bodies, surface mines, quarries, gravel pits, and urban areas, and 40 acres for all other categories. These minimum mapping units are the smallest areas appropriate for portrayal on 1:250,000-scale maps.

Acquisition of LU/LC digital data consists of the conversion of annotated graphic map detail to digital form: node positions to X-Y coordinate pairs, arc alignments into strings of X-Y values, and polygon category classifications into their digital representations. Manual "blind" X-Y digitizing systems, interactive digitizing/editing systems, and automatic laser line-following systems have been used by the USGS to provide raw digitized data to GIRAS. To date, approximately 250 digital LU/LC files and over 1,000 associated files reside in the NDCDB.

DIGITAL LINE GRAPHS (1:2,000,000)

Until this time, no suitable data base at a regional or national scale was available to users. The USGS, in response to increasing user requirements for up-to-date cartographic data and products on a nationwide basis, has introduced the 1:2,000,000-scale digital cartographic data base. The 1:2,000,000-scale general reference maps of The National Atlas of the United States of America served as the cartographic source materials for the digitizing. The categories of information in this data base include political boundaries (State and county level), administrative boundaries, transportation networks (roads and railroads), and hydrographic features (streams and water bodies). Just prior to the digitizing operations (1979-1982), the source materials were updated to assure that the most detailed and comprehensive information available was incorporated into the files. As a general indication of the higher level of information content and greater digitizing accuracy, the 1:2,000,000-scale data base contains some 15 million points to define the features of the United States. Previous data files such as World Data Bank I and World Data Bank II utilized 20,000 and 1.5 million points, respectively, to cover the world.

Digital Line Graph procedures, like those used to collect 1:24,000-scale DLG data, were employed in the capture of data for this data base. To reiterate, the data are digitized in a topologically complete, node-area-line structure; identifying beginning and ending nodes with areas left and right of a given line. In addition to the topological coding, attribute codes are applied to the features to provide additional information about the graph elements in the file. A feature code consists of two parts: a three-digit major code and a four-digit minor code. A major code of 290 is generally applied to a feature in the 1:2,000,000-scale data base to indicate that it is small-scale data base information. The four-digit minor code describes the feature. Coding of map features in this data base has been designed to facilitate the use of the data to produce small-scale maps, that is, 1:2,000,000- to 1:10,000,000-scale. Two built-in capabilities have been included to achieve this: (1) a generalization capability to allow exclusion of features as scale or intended use dictate (length or size of feature), and (2) a grouping capability to allow the formation of logical sets of information. As scales decrease, smaller sized features will be deleted and not made available for graphic presentation, contributing significantly to cartographic cleanliness. The continuity of linear features consisting of segments of varying attribute codes is an important application of the grouping capability.

The 1:2,000,000-scale DLG data are not only available in the standard topologically structured DLG format, but also in a graphics format that is suitable for plotting using the GS-CAM (Cartographic Automated Mapping) package. The graphic format identifies only the line descriptions and is derived from the standard DLG format. Minimal descriptive information is provided for each line and the relationships between and among lines is unspecified. Twenty-one files covering the United States are available from the NDCDB: 15 for the conterminous U.S., 5 for Alaska, and 1 for Hawaii.

DIGITAL LINE GRAPHS (1:100,000)

In April 1983, a Memorandum of Understanding signed by the USGS and the Bureau of the Census initiated the production of DLGs captured from the forty-eight 1:100,000-scale topographic base maps covering the State of Florida. Having established this proof of concept, the project was increased in scope to encompass the entire continental United States. In order for the Bureau of the Census to utilize the data obtained from this massive effort in their 1990 Decennial Census of Population and Housing, the project must be completed by 1987. The DLGs generated from this cooperative effort, the hydrography, the miscellaneous transportation, and the roads and trails overlays, will then take residence in the 1:100,000-scale DLG portion of the NDCDB. Parameters for the distribution of 1:100,000-scale DLG data are currently being identified by the USGS, and sales from the NDCDB should begin in mid-to-late 1985.

The Census Project, as it has come to be known, requires the DLG digitizing and processing of 1,807 1:100,000-scale quadrangles. This activity far exceeds NMD production capabilities using manual data capture methods. Consequently, the NMD has adopted a slightly more automated alternative to pure manual digitizing. The SciTex raster scanning of the "image" of the data layer in question, followed by the sequential editing, raster-to-vector conversion, structuring and attribute tagging, and UCLGES processing is providing a viable solution to the manual digitizing bottleneck. Another alternative being investigated is the automated line-following of data from feature separates.

DIGITAL ELEVATION MODELS (1:24,000)

US GeoData DEMs corresponding to the USGS 7.5-minute quadrangles are also being produced for the NDCDB. A DEM is defined as a data file consisting of a sampled array of elevations for a number of ground positions--a digital representation of the terrain surface. A square grid, with 30-meter cell increments, is digitally superimposed on the terrain surface defined by 7.5- by 7.5-minute quadrangle boundaries. Each intersection point within that 30-meter matrix represents a corresponding point (same X-Y coordinates) on the ground. The elevation of that ground point then becomes the Z-component of the coordinate triplet at that intersection. A DEM can be placed in either a 7-meter or 15-meter root mean square error (RMSE) data base depending on the vertical accuracy of the data.

A DEM file is of a raster structure, accommodating data acquired from many different sources. Capabilities exist within the USGS to create a DEM from photogrammetric stereomodel terrain profiling or contour digitizing, from the digital output of automatic image correlation orthophoto instrumentation, or the scanning or tracing of contour plates. Like the DLGs, DEMs may be classified at one of three levels depending on editing, enhancements, and spatial structuring.

DEM-1. DEM level 1 contains a network of raw elevation data that has only been edited for gross blunders and has not been keyed to the planimetry. The DEMs being data based by the USGS at present are at the DEM-1 level.

DEM-2. DEM level 2 contains elevation data that have been smoothed for consistency, enhanced to remove noise, and filtered to reduce data volume. The data have not been keyed to the planimetry.

DEM-3. DEM level 3 contains elevation data that have been edited and modified to be consistent with the planimetric features represented by a DLG-3 file.

Generally, the principal source of data for the 7.5-minute DEMs is high-altitude photography (nominal scale 1:80,000) flown in such a way that one photograph covers an entire 7.5-minute quadrangle (quad-centered). The quad-centered photograph is the common photo in the two stereomodels that cover the quad. Manual profiling of stereomodels provides raw digital data for a large number of DEMs being stored in the NDCDB. After a controlled stereomodel is oriented in a stereoplotter such as the Wild B-8, the tri-axis locator system is locked in the X-axis (along the flightline). The measuring mark is then mechanically driven in the Y-axis (perpendicular to flightline) at a constant or variable rate (operator controlled). The Z-axis (elevation of the model terrain surface) is controlled by the operator, who "keeps the measuring mark on the ground." All three axes in the stereoplotter are digitally encoded so that movements along each axis are recorded on tape. When the Y-axis profile has fully traversed the model, the measuring mark is stepped-over in the X-direction by a specified increment, and the second profile begins. This profile-stepover process continues until the entire model is completed. The resulting magnetic tape file contains raw terrain data for that model. Within the model space are aerotriangulated pass-points whose ground positions were determined during the block adjustment procedures. The model coordinates for these points are fit to the corresponding ground values using a three-dimensional transformation. The remainder of the profile points are then transformed to the ground system. Another software procedure then panels together the models comprising a quadrangle, resamples the data to a 30-meter elevation grid, and trims the data to quadrangle dimensions.

DEM data is also produced by the USGS using the Gestalt Photo-Mapper II (GPM-2) automatic image-correlation system. Density patterns (images) on analytically oriented stereopairs of diapositives are matched by the image correlation system, and parallaxes are measured and removed from the image patch, or measurement area. Since the parallaxes are directly related to the image heights, model X, Y, and Z coordinates are determined for each picture element in the patch. This procedure is performed until the entire stereomodel is digitized. In the GPM-2, the image patch is approximately 8 mm x 9 mm, generating 2,444 data points on a 182 micrometer grid. This translates to about 750,000 points for a typical quad-centered model. Like the profiled data, these data are processed through resampling routines that prepare 30-meter DEMs for entry into the NDCDB. In addition to raw digital data, the GPM-2 is capable of producing orthophoto imagery and contour graphics.

The Digital CArtographic Software System (DCASS) is another method employed by the USGS to prepare DEM data. DCASS was developed by the USGS to collect and utilize digital data obtained from a stereoplotter during map compilation. DCASS is a modular system designed to accommodate cartographic activities such as feature joining between stereomodels, batch or interactive editing, vector or raster graphical output, and the production of DEM data. To accomplish the DEM data production, a DCASS subsystem scans and grids compiled and elevation-tagged contour lines, using a two-axis interpolation algorithm. A custom-designed vector-to-raster process actually performs the conversion from digitized contour data to DEM format. North-south and then east-west scan line intersections with the contour lines are successively computed and resampled to 30-meter intervals. A merging of east-west with north-south scan-lines determines the elevation values for the 30-meter grid intersection. Further software processing prepares the DEM for data basing.

The optical scanning of contour plates (like the planimetric separates scanned for the 1:100,000-scale DLGs) provides an additional source for DEM data. The SciTex System utilizes a drum-type raster scanner capable of digitizing sheets up to 36 inches square at resolutions ranging between 4 and 47 points per millimeter at 130 revolutions per minute. A color design station on-line with a central processor is used to edit the data. The scanning of a contour plate produces a raster file that presents that graphic as a matrix of one's and zeroes. The raster-to-vector process that is necessary to accomplish the conversion from scan data to DEM file begins with a procedure to break down the massive volume of bit data into workable blocks (512 pixels square). The data within each block are then skeletonized. A contour line that was originally 2 or 4 pixels wide is reduced to a single pixel width. The resulting files are vectorized and merged with the other blocks representing the quadrangle to make a continuous data file. The vectorized data file is now suitable for interactive editing to tag the contour lines with elevation values. Resampling software then grids the contour data to the NDCDB compatible 30-meter intervals.

To give you an idea as to data volumes in 30-meter DEM files, a DEM in northern Washington contains about 141,000 elevation points. A DEM in southern Texas contains nearly 195,000 data points. Approximately 7,000 DEM files reside in the 7-meter data base, while another 3,000 DEMs are located in the 15-meter RMSE data base.

DEFENSE MAPPING AGENCY DIGITAL TERRAIN MODELS

US Geodata DTMs are digital records for regularly spaced points on the ground surface. Approximately 1,480 1° x 1° (latitude by longitude) blocks are available from the NDCDB. These data files are a byproduct of the Defense Mapping Agency Topographic Center (DMATC) program to produce plastic raised-relief maps. DMATC developed the Digital Graphics Recorder (DGR) in the early 1960's in an effort to streamline their mapping operations. The DGR system components included an automatic X-Y digitizing table online with a computer system. Contour lines were traced and converted to a matrix of terrain elevations by grid intersection and interpolation algorithms. The output from the DGR operation was used to drive milling machines that carved the plaster forms used to mold plastic raised-relief maps. Clearly, this gridded elevation data had tremendous potential outside the milling-machine controller business, due primarily to the computer-compatibility of the digital data. A vast number of graphical and analytical operations could be performed using these computerized data.

The original map materials from which the DTM data were acquired were the 1° x 2° 1:250,000-scale topographic quadrangles. DMATC had prepared two one-degree matrices to cover each quadrangle, labeled east and west, due to the large amount of data in a full quad sheet. DTMs are available in two formats: PLANAR and arc-second. PLANAR data exist in an X-Y system with elevation values at an interval of 0.01 inch at 1:250,000 scale or about 208 feet. At 40° North latitude, a 1° x 1° quad will be approximately 69 miles by 54 miles. The DTM will then contain approximately 2,600,000 elevation points (1,820 x 1,420). The arc-second format presents elevation data at a latitude and longitude interval of 3 arc-seconds. A $1^{\circ} \times 1^{\circ}$ (3,600 arc-seconds by 3,600 arc-seconds) quadrangle will contain 1,201 x 1,201 or 1,442,401 elevation points.

The existence of terrain data in digital form permits the computation of slope and aspect--two terrain variables that are frequently integrated with other earth science information for various applications. Shaded relief graphics based on sun reflectance and ground-slope computed from a DTM, when combined with other information such as LANDSAT or map overlay data, make effective cartographic products. Surface modeling and analysis, and other non-graphical applications for large-area studies are served well by DTM information.

OTHER DATA TYPES

Geographic names information is a US GeoData product also available from the National Cartographic Information Center. Digital geographic names data are collected and archived using a multipurpose software and data base complex called the Geographic Names Information System (GNIS). GNIS includes five distinct data bases:

- the National Geographic Names Data Base
- the USGS Topographic Map Names Data Base
- the Generic Data Base
- the National Atlas Data Base
- the Board on Geographic Names Data Base

Access to the various data bases is provided by a software package known as the General Information Processing SYstem (GIPSY), which provides search and sort routines that allow a user to selectively extract and assemble data for desired applications.

CONCLUSION

The information revolution will undoubtedly lead to the creation of more and larger data bases with a corresponding supporting cast of hardware and software systems for their use. The motivation in automating the data collection and analysis process was to use the tremendous power of the computer to ease the burden of integrating and synthesizing the large data sets involved in earth science and natural resource studies. If the individual data sets or data bases are developed independently, in isolation, problems of compatibility most certainly arise. Since the inherent usefulness of any information set should be its applicability to other sets of data, it is important that the ability to integrate these diverse data layers be retained. Information interchange, such as is being conducted at seminars like this one, will surely guarantee that later information interchange, in the form of data tapes, will not be hindered.

REFERENCES

1. Allder, William R., and Elassal, Atef A., 1983, Digital Line Graphs from 1:24,000-Scale Maps. U.S. Geological Survey Circular 895-C (USGS Digital Cartographic Data Standards).

2. Allder, William R., Szeide, A. Joan, McEwen, Robert B., and Beck, Francis J., 1983, Digital Line Graph Attribute Coding Standards. U.S. Geological Survey Circular 895-G (USGS Digital Cartographic Data Standards).

3. Callahan, George M., and Broome, Frederick R., 1984, The Joint Development of a National 1:100,000-Scale Digital Cartographic Data Base. 44th Annual Meeting of the American Congress on Surveying and Mapping, Washington, D.C., March 11-16, 1984, Technical Papers, p. 246-253.

4. Domaratz, Michael A., Hallam, Cheryl A., Schmidt, Warren E., and Calkins, Hugh W., 1983, Digital Line Graphs from 1:2,000,000-Scale Maps. U.S. Geological Survey Circular 895-D (USGS Digital Cartographic Data Standards).

5. Elassal, Atef A., and Caruso, Vincent M., 1983, Digital Elevation Models. U.S. Geological Survey Circular 895-B (USGS Digital Cartographic Data Standards).

6. Fegeas, Robin G., Claire, Robert W., Guptill, Stephen C., Anderson, K. Eric, and Hallam, Cheryl A., 1983, Land Use and Land Cover Digital Data. U.S. Geological Survey Circular 895-E (USGS Digital Cartographic Data Standards).

7. Frederick, Doyle G., and Anderson, K. Eric, 1982, The Collection and Analysis of Natural Resource Data in the 1980's. Fifth International Symposium on Computer-Assisted Cartography (Auto-Carto V), Crystal City, VA, August 22-28, 1982, Proceedings, p. 33-40.

8. McEwen, Robert B., Calkins, Hugh W., and Ramey, Benjamin S., 1983, Overview and USGS Activities. U.S. Geological Survey Circular 895-A (USGS Digital Cartographic Data Standards).

9. McEwen, Robert B., and Starr, Lowell E., 1984, Research for the USGS Digital Cartography Program. 1984 ACSM-ASP Fall Convention, San Antonio, TX, September 9-14, 1984, Technical Papers, p. 325-333.

10. McEwen, Robert B., 1983, Digital Cartography in the U.S. Federal Government. Harvard Computer Graphics Conference, Cambridge, MA, July 31-August 4, 1983, Proceedings, 27 p.

11. Payne, Roger L., 1983, Geographic Names Information System. U.S. Geological Survey Circular 895-F (USGS Digital Cartographic Data Standards).

12. Southard, Rupert B., and Anderson, K. Eric, 1982, A National Program

for Digital Cartography. Fifth International Symposium on Computer-Assisted Cartography (Auto-Carto V), Crystal City, VA, August 22-28, 1982, Proceedings, p. 41-49.

13. Starr, Lowell E., and Anderson, K. Eric, 1981, Some Thoughts on Cartographic and Geographic Information Systems for the 1980's. William T. Pecora Symposium, 7th, Sioux Falls, SD, October 18-21, 1981, Proceedings, p. 41-55.

BIOGRAPHICAL SKETCH

Alan M. Mikuni is Chief of the Digital Section at the U.S. Geological Survey National Mapping Division's Western Mapping Center in Menlo Park, California. He is responsible for the development of digital cartographic data sets suitable for entry into the National Digital Cartographic Data Base. He graduated from California State University, Fresno, with a B.S. in civil engineering in 1970 and has been with the U.S. Geological Survey since 1966. He is affiliated with a number of surveying, mapping, and engineering societies, including the American Society of Civil Engineers, American Society for Photogrammetry and Remote Sensing, American Congress on Surveying and Mapping, and National Society of Professional Engineers, and is a Registered Civil Engineer in California.

18

AUTOMATIC SCANNING AND DATA PROCESSING FOR CAD AND ARCHIVAL SYSTEMS

William Pferd, III
Vice President, Technology
Skantek Corporation
150 Mount Bethel Road
Warren, New Jersey 07060

Abstract

The benefits from design automation, increasingly evident through the pervasive growth of CAD, have been achieved to date almost exclusively by keyboard and manual digitizing data entry methods. During the last few years, some entry of data from the vast number of existing engineering drawings has been demonstrated by automatic means through advances in optics, electronics, microcomputers and artificial intelligence. Now, these previously proven laboratory methods are being embodied in an increasing variety of automatic data capture machinery. In this paper, the more desirable features of automatic data capture machinery are discussed along with the variety of elements that can be used to fulfill the functions provided by the scanning and data processing sub systems. Also included is a description of the Skantek Corporation SK-1010 Automatic Digitizer, and it is shown how this new machine provides improvements in the automatic capture of engineering drawings. The paper concludes with the presentation of hardcopy reproductions of files created after scanning a typical mechanical part drawing on the SK-1010 along with compression and CAD entity statistics for the processed data.

1. INTRODUCTION

Although computer aided design (CAD) systems have been in use for over ten years, there are still mountains of manually prepared drawings in the world today. In fact, recent studies show that there are many millions of active drawings in the U.S. alone. Most companies file thousands of drawings in their vaults, with approximately 20 percent active each year. The manual drafting effort that is required for these drawings continues to be sizeable. Additionally, it is estimated that 10 percent of the drafting effort in the U.S. is expended merely in drawing maintenance. During use, they become dirty, smudged or torn and must be refurbished regularly. The long standing problem has been how to get these drawings into computer-based systems for automated production, storage, retrieval and display of the engineering data.

Installing a CAD system is a major decision that involves not only cost factors, but the problem of how to manage the capture of data from older drawings and new ones that continue to be prepared in-house or that are received from vendors. Today, a number of new venture companies recognize this problem and are engaged in bringing new solutions to the CAD marketplace. This paper presents the goals for automatic data capture and the performance characteristics of various scanning-processing machine elements. Also described are the features of a new type of Automatic Drawing Capture Machine (ADCM) that is designed specifically to capture data from the largest of engineering drawings and serve as a peripheral for CAD and archival systems.

2. DRAWINGS FOR CAPTURE

The major types of drawings of interest for automatic data capture and conversion are listed in Table I. These are the kinds of documents that are invaluable to industry and government. In general, engineering documents are either non-scaled or scaled and have a few or many dimensions or alphanumerics. The non-scaled, non-dimensioned types have been shown to require little processing to gain CAD editing utility. The conversion of scaled and fully dimensioned types of drawings is a more formidable task and requires the use of artificial intelligence methods and some operator intervention to duplicate an exact CAD file. An acceptable ADCM should be able to handle this wide variety of engineering documents that are encountered in design, manufacture, construction, installation and repair of equipment and facilities.

TABLE I - TYPES OF DRAWINGS THAT NEED CAPTURE AND CONVERSION

1. Non-Scaled, Non-Dimensioned
   - Electrical Wiring Diagrams
   - Technical Illustrations
   - Field Assembly Drawings
   - Architectural Drawings
   - Piping and Instrumentation Diagrams

2. Non-Scaled, Few Dimensions
   - Piping Isometrics
   - Plant/Process Layouts
   - Specification Sheets

3. Scaled, Few Dimensions
   - Electronic/Electrical Schematics
   - Facility Drawings
   - Structural Drawings
   - Lofting/Template Drawings
   - Mechanical Assemblies
   - Area/Utility Maps
   - Site Plans

4. Scaled, Many Dimensions
   - Mechanical Details for Manufacture
   - Shop Subassembly Drawings
   - Non-Isometric Assembly Drawings

The dimensions for engineering drawings defined in American and International Standards must be used to set the requirements for the scanner. These standards show that, to be able to process drawings used by the major industrial sectors, the ADCM must be able to handle drawings up to

1 meter wide and having a length of up to 3 meters. Because of minimum width pencil or pen marks on these drawings, very finely spaced data acquisition sites should be provided in order to permit complete recognition of the graphic and text information. A minimum of three bits of data per minimum line width is desirable. It is apparent that a very large amount of data measured in millions of bits per drawing will be generated and that the use of very efficient systems for processing the data will be required.

3. GOALS FOR AUTOMATIC DRAWING CAPTURE MACHINES

Most older ADCM's have evolved from designs that were developed earlier for use in the publishing or cartographic fields. These machines, not bound by the need to operate in an engineering office environment, employ large computers, are awkward to operate, yet have highly accurate optical scanning systems. They anticipate use at a centralized facility by closely supervised and specially trained operators. Until recently, there has been no automatic data capture machine that was designed and developed especially for use in an engineering drafting office and as a convenience device. For this type of use, entirely new machine concepts are required to place the ADCM in the category of a CAD peripheral such as the electrostatic plotter or high resolution workstation.

The more important design and performance goals for a large document ADCM are listed in Table II. First, it must be reliable and simple to use. It should respond quickly to simple instructions in handling the variety of engineering documentation. The machine should be "user friendly" and "intelligent" with automatic controls for the capture and subsequent conversion of noise free data. A compact shape of desk size, with low maintenance to match that found in other modern office machines is needed. Standard formatted files defined in CCITT (facsimile) standards and by the Initial Graphics Exchange Specification (IGES) should be created along with other formats that are targeted to the major CAD systems currently in use worldwide. Finally, the product should employ

TABLE II - DESIRABLE DESIGN AND PERFORMANCE FEATURES FOR LARGE DOCUMENT AUTOMATIC CAPTURE MACHINES

- RELIABLE - employs high performance and easy to maintain optical devices
- EASY-TO-USE - touch operation with software controlled instructions
- FLEXIBLE - handles documents up to 1 meter wide and 3 meters long
- SIMPLE - small desk-top unit with straight-through document transport
- COMPATIBLE - CCITT, IGES and native CAD output encoding
- STAND-ALONE - integrated microprocessing to control all machine functions
- EXPANDABLE - to provide automatic recognition of characters, symbols and geometric elements; to provide accurate CAD type geometry
- EASE OF INSTALLATION - no special environment or power requirements
- PRICE/PERFORMANCE - payback in less than 12 months

modern optics and microelectronic technology in order to be attractively priced and give high performance.

These goals can be considered aggressive and difficult to achieve, yet advances during the past few years in such diverse fields as optical fibers, charge-coupled devices, microprocessors, color monitors, touch-screen controls, pattern recognition and image processing now make their attainment possible.

4. FUNCTIONALITY FOR ADCM

Figure 1 identifies the more important hardware and software functions that must be provided by the elements that comprise an ADCM.(1) As shown on the left of Figure 1, the role of the Optical Scanner Sub-system is to direct light from the source of illumination to a light detector by way of the document. The different types of elements that can be used for the illuminator, aperture, scanner and detector are listed in Table III. The choice of elements influences significantly the size, performance, handiness and cost of the sub-system.

Optical Scanner Sub-System			Data Processing Sub-System		
	Drawing	Capture	Machine	Control	
Document Illuminator	Document Transporter Holder	Light Detector	Analog to Digital Converter	Data Condition-ing	Data Encoder
	Sampling Aperture or Scan Mechanism			Digital Data File	

FIGURE 1 - OPTICAL SCANNER AND DATA PROCESSING SUB-SYSTEMS FUNCTIONALITY

Large document scanners impose special considerations on the choice of elements. The need to accommodate Standard "E" and "K" sizes effectively eliminates many of the options since the assembly would become too large or the light path excessively long in order to acquire data from the wide documents. Shorter light paths can be obtained by using a number of linear electronic detectors, each focused through lenses on a portion of the line width to be scanned. However, this technique imposes critical alignment requirements, imposes severe demands on the illuminator and results in bulky configurations. An elegant combination for large document scanners that offers the smallest size and greatest ruggedness is lamp illuminated optical fibers (the aperture) married to an array solid state imaging device (the detector). This combination functions as a hybrid flying spot/aperture type of scanner with one row of optical fibers performing spot illumination and a second row coupled to the imaging device to perform spot detection.

Optical fibers can provide a number of other inherent advantages besides compact size and stability. The flexibility of fibers allows great freedom in locating the illuminator and detector. Therefore, the manner in which the document feeds into the machine can be given primary consideration and need not be compromised by the positioning of the optical elements. For example, the document can be positioned with the information side visible to the machine operator at all times. Finally, the high resolution required to scan fine lines can be easily achieved with optical fibers and at an attractive cost.

TABLE III - POSSIBLE ELEMENTS FOR OPTICAL SCANNER SUB-SYSTEMS

Illuminator	. Laser . LED . Lamp . CRT	Scanner	. Drum . Flatbed . Mirror . Roll Feed
Aperture	. Acousto-Optic . Electron Beam . Optical Fiber	Detector	. Vidicon . CCD . CID . CCPD

The ADCM Data Processing Sub-system, as shown on the right side of Figure 1, comprises data conversion, conditioning, encoding and file management. The choice of elements for these functions can involve both hardware and software considerations. Scanning of the original document resolves the line and alphanumeric image into small picture elements or pels. At the converter stage, pel output data can be either 8-bit (gray scale) or thresholded (binary). If required, the data can be moved directly to the file manager as a raster file. Alternatively, a conditioning function can be used to improve the quality of the scanned data or remove noise from the file that results from the scanning of dirt or stray marks on the document. Various algorithms in hardware that process rapidly the binary data, with or without intermediate storage, are available for these purposes. Typically, the data from the conditioning process is filtered and run length coded (RLC), similar to facsimile data. This type of data can be sent as an image file over transmission paths for local or remote display on a raster terminal, output directly on a plotter or placed in file for archival purposes.

The last Data Processing Sub-system function, encoding, is critical for the use of an ADCM in CAD. Ideally, the encoding further converts the RLC data into a compact form that can be readily accepted by the CAD system software. Encoding must be an optimally designed process due to the enormous quantities of data involved, e.g., the thresholded bit stream from an "E" size drawing digitized at 10 pels per millimeter in orthogonal directions will contain upwards of 100 million binary bits. Considering this factor and also the conversion cost, speed objective, and the ultimate use of drawing image files in CAD, the encoding methods employed are a continuing subject of concern and innovative development.

In general, the methods can be characterized as those that do or do not require intermediate storage of the conditioned data before encoding and those that encode the data into geometric entities by edge, area or structural determinations that apply artificial intelligence types of algorithms. All methods aim to miminize both the time to process the data, the data count in the CAD compatible file and the post-encoding editing time in the CAD system during changes and rearrangements of the file. The most desirable results of encoding would include, in order of processing: blobs of compressed RLC, solid and dashed line entities, alphanumerics in lines or ASCII Code, text strings and, finally, graphic entities of chained vectors, circular arcs, conics, splines, arrows and point dimensions. The encoding of the graphic entities offers distinct data compression and editing advantages for the CAD compatible file. It can be expected that encoding methods will eventually provide files containing all these levels of refinement. However, currently available ADCM's, while encoding only RLC or a limited set of geometric entity files, can provide benefits today to justify their use in CAD and archival systems.

The ADCM Control function, shown across the top in Figure 1, should ideally be involved with all the other functions of the Optical Scanner and Image Processing Sub-systems. The design choices for the control functions determine the ease-of-use, the self-correcting nature and the degree of automatic performance. Examples for intelligent controls are scanning area determination, contrast and illumination settings and providing information about the state and results of scanning and processing. In order to reach many of the Table II performance goals, special attention must also be given to the type of machine interfaces presented to the user. The controls should facilitate the human interaction by addressing important physiological and cognitive aspects. Proximity for easy tactile action, color to highlight priority conditions and care in signing the controls to aid comprehension are some of the examples of human factors consideration.

5. A NEW ADCM PERIPHERAL

The SK-1010 Automatic Digitizer, developed by the Skantek Corporation for automatic capture and conversion of engineering drawings, is shown in Figure 2.(2) It is composed of four integrated sub-systems that are engineered to meet the design and performance goals shown in Table II. These sub-systems are the Scanner, Electronic, Magnetic Tape and Monitor Units. The Scanner Unit performs the automatic digitization and is equivalent in functional behavior to the Optical Scanner Sub-system shown in Figure 1. The Electronic Unit performs the Data Processor Sub-system functions and contains a microcomputer, hard and floppy disk drives, various software modules and output connections. The Magnetic Tape Unit facilitates transfer and off-line storage of the data. The Monitor Unit provides controls and instructions for the operator through an intelligent touch screen and also displays the digitized data that results after the analog to digital conversion. The Units are packaged separately and interconnected with cables on a unit bench at the customers site to form a stand-alone assembly.

The SK-1010 operates in a normal office environment. The microcomputer and data conversion software process the scanned data into a variety of forms. Eleven identifiable major processes with the SK-1010 are shown on Figure 3, along with the currently available and planned software products. The starred items indicate the processes that are furnished with the SK-1010. Processes 3 and 4, Vectorizing and Entity Recognition, includes currently software for conversion of raster data to connected lines, arrowheads and circles. Future additions to the Entity Process will include recognition of arcs, splines, dashed lines, point dimensions and certain non-geometric entities. The Text Recognition Process will

FIGURE 2 - THE SK-1010 AUTOMATIC DIGITIZER

encode to ASCII the letters and numbers on documents in hand drawn, LeRoy and Veritype forms. The SK-1010 and its standard and optional software packages are designed to produce a CAD compatible data file to the IGES specification that contains:

- Precisely reproduced and accurate raster and layered entity-based data
- The more common geometric elements represented in CAD (IGES) format
- Common text fonts converted to American Standard Code for Information Interchange (ASCII) notation
- Structural associativity and connectivity information
- Recognized symbols along with associated text and nodel data

6. SK-1010 AUTOMATIC DIGITIZER ARCHITECTURE

The major elements and their interconnections in the SK-1010 are shown schematically in Figure 4. The optical fiber assembly in the unique Scanner Unit permits scanning with high resolution and precision (3). A solid state electronic array in the CCD Camera generates the digital signals that are processed into compressed data files in the central processor. This processing involves the conversion of 8-bit data to RLC data or into vector code that is further compressed by recognition software. A high resolution color CRT is used to preview the scanned file and display control information for touch screen interaction by the SK-1010 operator. Transfer to the CAD system is accomplished through magnetic tape.

SK-1010 PROCESSES	SOFTWARE PRODUCTS
DATA PROCESSING SUB-SYSTEM ELECTRONIC AND TAPE UNITS	
1 - 8-Bit Raster Data	
2 - Binary Raster Data	*
3 - Vectorizing	*
4 - Entity Recognition	*
5 - Character Recognition	SK-401 Text Package (OCR)(2Q85)
6 - Symbol Recognition	SK-402 Symbol Package (1985)
7 - Layering	SK-301 Layer Separation Package
8 - Formatting	IGES Interface* SK-102 CADAM Interface Package SK-103 CALMA (CEF) Interface Package (Others planned)
9 - Transfer	IGES Formatted Data* Run-Length-Code Data*
10 - Store	1600 BPI Magnetic Tape*

*Standard with SK-1010 Automatic Digitizer

FIGURE 3 - STAGES OF THE SK-1010 PROCESSING OF SCANNED DRAWINGS AND RELATED SKANTEK SOFTWARE PRODUCTS

Optical Scanner Sub-System

For long prints that are usually rolled when stored or carried, an in-feed tray is provided at the front of the Scanner Unit for holding the document prior to insertion into the horizontal input slot that extends across the scanner. Smaller drawings are inserted directly into the open Scanner Unit as shown in Figure 2. The broad, flat surface in front of the slot facilitates the insertion and positioning of the drawing. The document feed mechanism is normally open so that the drawing is easily inserted to a distance of a few inches. No special care is required since the roller feed mechanism automatically closes and positions the drawing to the top edge or user defined location when scanning is requested. The exact positioning is achieved by means of optical sensing. The drawing is backed out of the unit until the desired start point is just at the scan line of the optical fiber assembly. Between each scan, the document is advanced in small increments. Ultimately, the entire drawing is moved into the out feed tray where it is contained by a hinged cover until removed by the operator. Stiff board-like materials can also be scanned. A rear flap, when opened, allows a straight path through the unit.

Data Processing Sub-system

The data processing circuits for the SK-1010 are included in the Electronic and Tape Units. A microcomputer running the UNIX operating system is used to control all processes.(4) The microcomputer responds to commands that are received on activation of the resistive membrane touchscreen on the CRT. The camera controller and video processor in the Electronic Unit, indicated in Figure 4, provides conversion of gray level scan data to binary form. Further conditioning of the signals is performed in an I/O processor that also outputs run-length-coded (RLC) data to the microprocessor Multibus via direct memory access.(5) Special SK-1010 software resides in the microcomputer for encoding and formatting the data that can be either passed along the Multibus to the graphics display controller to paint the color CRT or to the magnetic tape controller for recording with the magnetic tape drive. Ample random access and hard disk memory is provided to support the operating system and data processing programs and to store the CAD compatible files that are generated. A diskette is provided to load the system programs for field servicing and to introduce future enhancements and special software routines.

7. SK-1010 AUTOMATIC DIGITIZER PERFORMANCE

Ideally, full scale plots or a high resolution CAD terminal should be used to evaluate the results of the processing of documents in a ADCM. An unavoidable alternative here is to present reproductions of plots of representative files. Included here are reduced reproductions of files generated on the SK-1010 that were recorded on a 10 nib per millimeter electro-static plotter with a 200 micron dot size. The collection demonstrates the performance obtained when capturing the scaled, dimensioned mechanical type of drawing listed in Table I.

Figure 5a shows the SK-1010 raster data capture (10 pels per millimeter obtained with optical fibers) for a typical D size mechanical part drawing. Figure 5b shows the graphics border layer. This, and the subsequent figures, are obtained with the SK-301 Layer Separation Package. Figure 5c shows the text layer as connected lines. The clearly legible small letters and numbers show the encoding accuracy. The alphanumerics, also available in raster format, can serve as the basis for character recognition to create an ASCII coded alphanumeric layer.

A plot of the IGES graphic layer for CAD terminal editing in connected line form is shown in Figure 5d. Note that the circles and the arrows on orthogonal axes are recognized. Also, up to 8 line weights can be selected by the operator for additional layered plots of the design graphics. Anticipated further enhancements through geometric entity recognition will result in near exact replication of originals and further reduced file size.

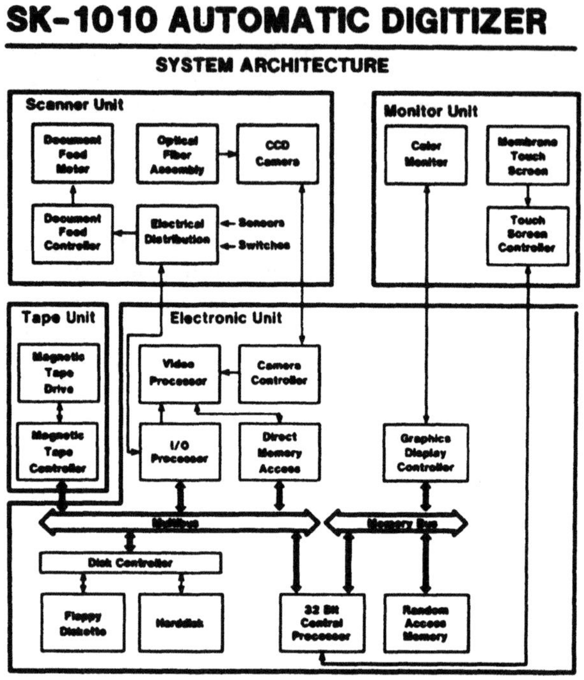

FIGURE 4 - SKANTEK SK-1010 AUTOMATIC DIGITIZER SYSTEM ARCHITECTURE

The files represented by the Figures are totally compatible with CAD and can be changed or rearranged in the same manner as if they were generated by keyboard commands on a CAD terminal. For the Figure 5 example, the processing times are but a few minutes. For "E" size drawings with average graphic and text content, the processing time to obtain similar results is proportionately longer. Some additional time is taken by the CAD workstation operator to perform acceptance checks and cleanup, where needed, of the scan created file. This usually involves the addition of some line entities and the substituting of geometric entities for selected line entity sets.

A portion of the IGES format file for Figures 5b, c and d is shown in Table IV. Note that the data is compressed into the IGES geometric entities of lines, copious data (multi-point pieces), circular arcs and leaders (arrows). The size of the files, in bytes of data, that result from ADCM scanning and processing effects the memory capacity requirements and response speed of the CAD system and, therefore, should be as small as possible. File size data for Figure 5 when in raster and scanner processed form are shown in Table V. An overall improvement of 25 to 1 is indicated from raster code to the CAD formatted compatible file. The amount of data is dependent on the Skantek algorithms, the amount of graphics and text on a drawing and the formatting overhead. Additional

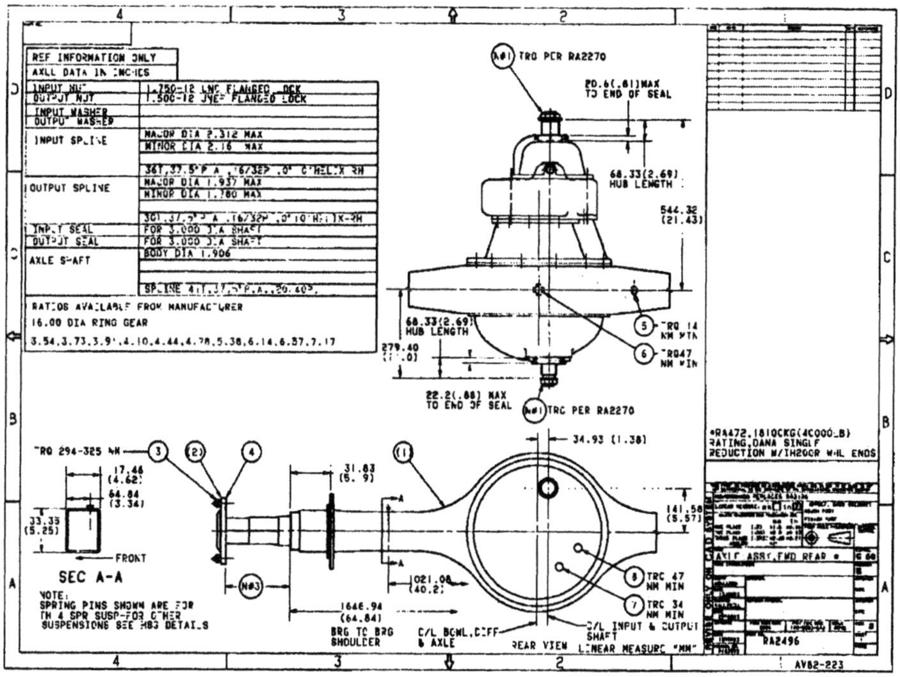

FIGURE 5a. - A REDUCED PLOT OF THE RASTER FILE OBTAINED AFTER THE SCAN OF A C-SIZE ORIGINAL WITH THE SK-1010

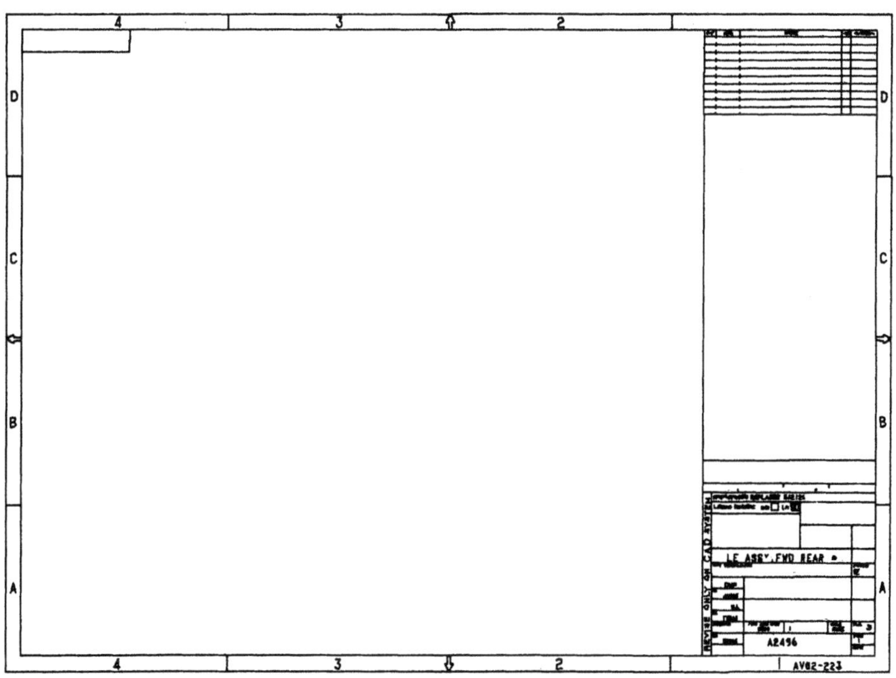

FIGURE 5b. - A PLOT OF THE BORDER FROM THE SK-301 LAYER PACKAGE

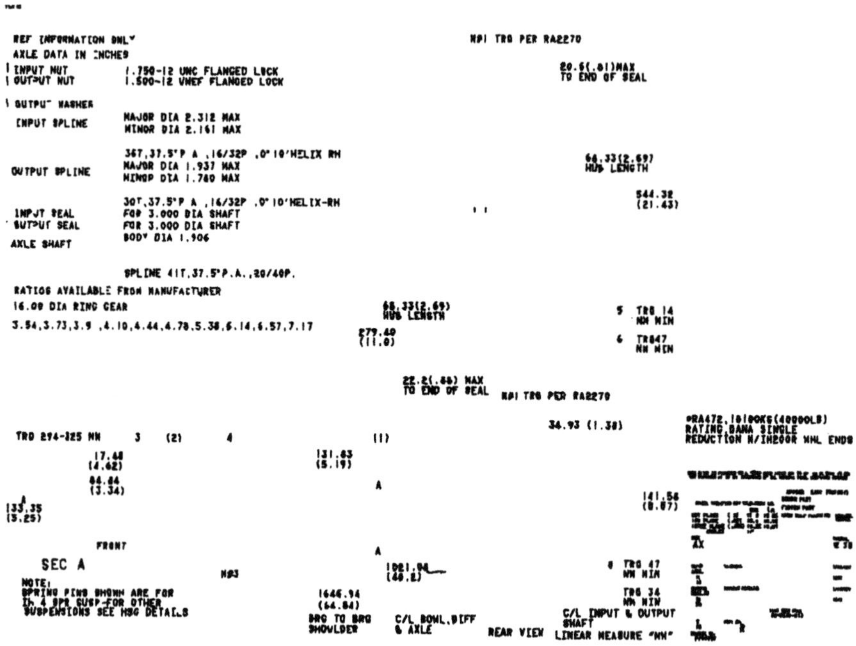

FIGURE 5c. - A PLOT OF THE TEXT LAYER IN CONNECTED LINE FORM

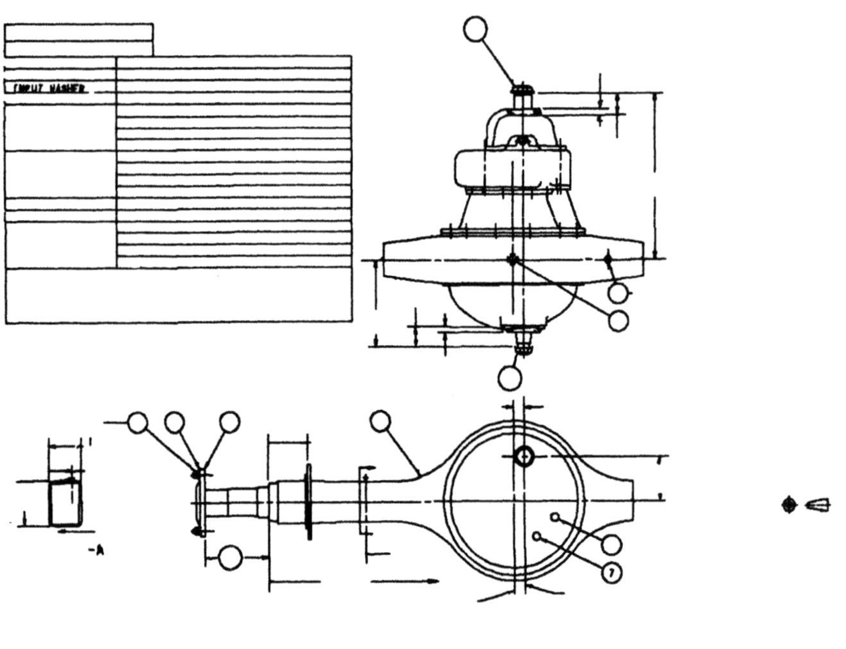

FIGURE 5d. - A PLOT OF THE GRAPHIC LAYER FROM THE SK-301 LAYER PACKAGE

compression of the data can result from the recognition and conversion of text to ASCII format. Next in effectiveness is the conversion of circles and arrows to CAD format. With the SK-1010 and the SK-401 Text Package, the resultant files will permit easy manipulation in the typical CAD system.

TABLE IV - A PORTION OF THE IGES FILE OF FIGURES 5b, c AND d FROM THE SK-1010 AUTOMATIC DIGITIZER

SKANTEK IGES FORMAT Version: IG-2.00
Skanstation No.: 5 Drawing ID: AXLE
Units: Inches Length: 17.217 Width: 24.016
Start: 0.000 Stop: 11.024 Right: 0.000 Left: 17.323
Bright: 5 Contrast: 6 Average Threshold: 100.00
Rotation Angle: 0
DRAWING: AXLE, 13H SKANTEK VR-2.00

Directory Entry Section:

110	1	1	1	1	0	000000	D	1
110	3	1	1	0		L	1D	2
106	2	1	1	1	0	000000	D	3
106	5	1	2	11		L	3D	4
110	4	1	1	1	0	000000	D	5
.	.	.	.	.	.	.	.	.
.	.	.	.	.	.	.	.	.

Parameter Data Section:

110*	1.093	-10.069	0.0	8.719	5.147,	0.0	1P	1
110	1.093	-10.060	0.0	-8.710	-5.147,	0.0	3P	2
.	.	.	.	.	.	.	.	.
.	.	.	.	.	.	.	.	.
106*	1	3	0.000	-6.539	-6.626	-6.524	8243P	4679
-6.650	-6.555	-6.701					8243P	4680
106	1	3	0.000	9.787	-4.776	9.850	12797P	7263
-4.831	9.874	-4.921					12797P	7264
.	.	.	.	.	.	.	.	.
.	.	.	.	.	.	.	.	.
214*	00002	0.059	0.161	0.000	-1.138	1.224	12677P	7181
-1.138	1.051	-1.138	-0.021				12677P	7182
214	00002	0.059	0.169	0.000	0.681	-0.465	12679P	7183
0.681	-0.283	0.681	-0.161				12679P	7154
.	.	.	.	.	.	.	.	.
.	.	.	.	.	.	.	.	.
100*	0.000	5.260	-0.335	5.008	-0.335	5.008	12695P	7199
-0.335							12695P	7200
100*	0.000	2.406	-1.776	2.106	-1.776	2.106	12697P	7201
-1.776							12697P	7202
.	.	.	.	.	.	.	.	.
.	.	.	.	.	.	.	.	.

*IGES Entities: 110-Line, 106-Copious Data, 214-Leader (Arrow), 100-Circular Arc

TABLE V - FILE SIZE COMPARISON OF FIGURE 5 FROM THE SK-1010 AUTOMATIC DIGITIZER AND SK-301 LAYER SEPARATION PACKAGE

File Type	Bytes	Normalized
Raster Code	3,290,000	25
Run-Length Code (Fig. 5a)	591,000	4.5
Border (With Alphanumerics) (Fig. 5b)	11,200	
Vector Text Layer (Fig. 5c)	60,600	1
Design Graphics Layer (Fig. 5d)	58,300	

8. CONCLUSIONS

Automatic data capture is now practical for getting drawings into computer-based systems for CAD and archival purposes and surpasses in performance older manual or semi-automatic methods. Users of the SK-1010 Automatic Digitizer report productivity gains of up to 4 times over CAD re-draw of original mechanical assembly drawings and up to 10 times over manual digitizing. Similar productivity gains are occurring for many of the other types of engineering documents listed in Table I. Experience shows that automating the capture and processing of geometric, dimensional and text data from engineering drawings and illustrations with the SK-1010 leads to sizeable increases in efficiency in CAD and archival system production.

REFERENCES

1- Nagy, G., May, 1983, "Optical Scanning Digitizers," Computer. pp. 13-24.

2- Pferd III, W., November, 1984, "The Skantek SK-1010 Automatic Digitizer - A Large Document Scanner, Image Processor and Formatter for CAD Systems," Proceedings of CAD/CAM USA '84 Exhibition, U.S. Export Development Office, Tokyo, Japan.

3- Molnar, M., Booth, T., August, 1984, "A Fiber Optic/CCD System for Automated Capture of Drawings." Proceedings of the Society of Photographic and Instrumentation Engineers.

4- UNIX is a trademark of AT&T Bell Laboratories.

5- MULTIBUS is a trademark of Intel Corporation.

BIOGRAPHY

William Pferd III is a Founder, Vice President of Technology and Director of Skantek Corporation. He originated and advanced the concepts that are the basis of the Skantek Corporation automatic data capture technology.

Prior to Skantek, Pferd was with Bell Telephone Laboratories for 35 years where he worked on telephone equipment, communication systems and computer systems and software. He has reported his work in many publications and has received several professional awards. Pferd has been granted 70 patents. He is a graduate of Rutgers University and has a Masters Degree in Mechanical Engineering from the New Jersey Institute of Technology. He is a Senior Member of IEEE and a member of ACM and SPIE.

19

THE ROLE OF HUMAN FACTORS IN INTRODUCING GRAPHIC ARTISTS TO COMPUTER GRAPHICS

Betsy A. Riley
Union Carbide Corporation
Nuclear Division
Oak Ridge, Tennessee 37830 U.S.A.

Abstract

A presentation- and publication-quality computer graphics system has been implemented at Union Carbide Corporation Nuclear Division over the past four years. Success of the implementation required close interaction between programmers and illustrators. This paper discusses the problems involved in establishing a computer graphics capability in a conventional graphic arts department. The problems dealt with fall into three areas: identifying and acquiring appropriate hardware, acquiring user-friendly software that could meet stringent quality standards, and overcoming the prejudices and misconceptions of all the people involved.

1. INTRODUCTION

The purpose of this paper is to describe the methods used by Union Carbide Corporation Nuclear Division (UCC-ND) in establishing a computer graphics capability in a conventional graphic arts department so that others may benefit from our experiences. We found that success of this project required careful attention to three potential problem areas: hardware, software, and human relations. The first two areas had obvious needs for planning since acquisition decisions had to be made. The objectives on which we based our decisions on hardware and software are included in this paper. Although

---

establishing these objectives was not simple, once the objectives were agreed upon we had little trouble in our acquisition of hardware and software. The problems we encountered were in an unexpected area--human relations. Because there were no decisions to be made in this area, we did not consider it as a potential problem and so we did no planning. That was a mistake. The causes of the problems we encountered and the corrective actions we determined are also listed in this paper.

The hardware and software decisions we made are better understood if background information about UCC-ND is explained. UCC-ND operates the Oak Ridge National Laboratory (ORNL) and two other production facilities in Oak Ridge as a contractor for the Department of Energy (DOE).* ORNL is a large research and development installation employing 5000. Because computers and computer graphics were established research tools at ORNL long before we began the project with graphic arts, much of the hardware and software we would need had already been acquired. User-friendly time-sharing computers were used for editing and for debugging runs and as job entry stations for powerful batch computers. A precision computer output microfilm (COM) recorder, an electrostatic plotter and a multi-pen drum plotter were available for graphics output. A device- and machine-independent graphics subroutine library had been acquired for both the batch and time-sharing computers. The existence of all of this equipment had a great impact on our decisions, in fact triggering the decision to bring computer graphics to the graphic arts department. Our reasons were twofold. Graphic arts was overloaded with work and they hoped that use of computer graphics could increase their productivity. Secondly, we wanted to do something about the large numbers of computer charts being redrawn because they were not publication quality, since we were sure the hardware and software we had was capable of publication quality output.

2. HARDWARE

With these goals in mind, we attacked the most obvious problem--hardware for the graphic arts department--and came up with the following set of objectives:

- o Final output must meet the quality standards applied to conventional (hand-drawn) artwork.
- o Turnaround time must be comparable to work done by conventional methods.
- o Cost of workstations must not be so high as to restrict growth.
- o A combination of interactive capability (for development) and batch processing (for production) was needed.
- o Capacity must be sufficient to handle projected volumes of work and expected increases.

While identifying these objectives was relatively easy, determining if they were satisfied was not. For instance, we discovered that resolution was not necessarily the most important factor in judging computer graphics output. Reproduction quality was very important. This was because the computer-generated plot was not necessarily the final output. The final output might

---

*Operated by Union Carbide Corporation under contract W-7405-eng-26 with the U.S. Department of Energy.

be a 3-inch column figure or a glossy photo mounted on matteboard in a poster session. To judge the quality of a device, we had to take its output through various photographic and reproduction steps and then compare it to conventional artwork that had gone through the same processes.

Applying this test to output from our existing graphics devices led to some surprises. Our highest resolution device, the COM recorder, made the poorest showing in the small journal figure test. The extremely fine lines produced on this device looked very impressive on photographic prints (glossies) but did not reproduce well, tending to break up or disappear altogether. Attempts to overcome this problem by lengthening the exposure time only served to blur the image. The second surprise was that our lowest resolution device, the electrostatic plotter, gave the best results. The most effective method was to produce the original plot several times larger than the desired final size using thickened lines and then to reduce it on an office copier. Use of an office copier was important since line thickness was retained better in the reduced copy than in photographic methods where reducing the image reduced the thickness of the lines until they disappeared. For the first year or two of the project, the electrostatic plotter was used for publication plots. However, as time passed, the heavy load on the plotter took its toll. The quality of the electrostatic plots began to deteriorate. Meanwhile, we had changed both the CRT and the photographic methods used with our COM device, resulting in greatly improved reproduction quality. Consequently, we now use our COM device for publication plots and high quality proofs and the electrostatic primarily for rough proof copies. While our multi-pen plotter could have been used for this output, the difference in turnaround time made use of the other plotters more desirable.

In fact, turnaround time was another area in which we had trouble comparing computer graphics to conventional artwork. Two separate factors had to be considered: the amount of time the illustrator spent producing the chart and the amount of time before the finished copy was ready to give to the customer. In presentation graphics (color slides), there was no contest. The process used to generate color slides manually was so lengthy and involved that the 24- to 36-hour turnaround associated with computer graphics was a welcome relief. In fact, the ease of producing color slides was the main attraction of computer graphics to most of the illustrators. They have found that color text slides can consistently be produced on the computer many times faster and cheaper than manually.

Publication graphics were another story. An ordinary bar chart might take a skilled illustrator 45 minutes to one hour to produce using manual drafting methods. The same chart could be produced with 10 to 15 minutes effort using the computer, but (and here lay the problem) there was a 4 to 6 hour minimum wait before the output was ready. This difference at first seemed to indicate use of a desk top plotter located in graphic arts rather than use of the higher quality batch devices. However, we had overlooked the fact that many of the jobs coming into graphic arts were groups of five to ten plots, and turnaround time to the customer was the time it took to do the entire job. When we used sets of plots for comparison of turnaround time, we found tha the computer could easily match the turnaround time of conventional methods while requiring much less of the illustrator's time. As the illustrators have gained expertise and the software has become more sophisticated, this performance has improved. The computer now allows savings in costs and time, even on single quantity bar charts.

The savings in time on large sets of charts made us realize that two different capabilities were indicated and that a local desk top plotter was not suitable. While interactive capability was needed for developing and viewing a prototype chart, large production runs were more efficiently done in batch. Much of the savings in illustrator time would be lost if each chart had to be viewed and plotted. Since many of the charts in a large job are very similar, the illustrators found that, after they set up a typical chart, they could modify the commands to produce the other charts in a much shorter time, without the necessity of viewing each individual chart. However, recent quality improvements and cost reductions in desk top plotters are causing us to consider the feasibilithy of desk tops for rush jobs.

The original considerations above led us to configure a developmental workstation consisting of a large screen high-resolution graphics terminal with a proof-quality hardcopy device and a digitizing tablet. (While this was the optimum workstation, for many applications a small screen storage tube terminal was found to be sufficient.) This equipment was connected to our user-friendly time-sharing computers. For production needs, a data entry or word processing terminal was sufficient to allow the illustrator to submit jobs to the batch computers to be plotted on the COM or electrostatic plotter. In practice, we have found that the tablet is used less than we expected, due to overloading of computing resources. Similarly, the graphics terminals are frequently used just to edit files, with illustrators submitting to batch for proof copies rather than previewing work.

3. SOFTWARE

Along with our consideration of hardware, we gave attention to the other obvious problem--what software should be used by the graphic arts department. Software had to be considered carefully since it could make or break the project. Software is also a continuing expense and over the life of a project can run several times the cost of the hardware. After considering our applications and our hardware choices, we generated a set of objectives for software.

- Output must meet the quality standards applied to conventional artwork.
- Software must be device- and machine-independent.
- The system must be user-friendly.
- Software must be flexible and capable of growth to meet future needs.
- Software must be compatible with existing corporate applications.
- Programmer support level required to develop and maintain the software should not be excessive.

Of these objectives, meeting publication quality standards with computer graphics was the toughest. It was this problem that led us to the graphic arts department in the beginning. We wanted to find out what was needed to achieve publication quality.

The illustrators identified several features that were needed and usually missing on computer-generated charts. Most of these turned out not to be due to inadequacies of the software but to ignorance on the part of the programmers. The main features we identified are as follows: Line weights on a chart should vary with the most important chart element (the data curve) having the thickest line. Sans serif fonts should be used (this was the most common failing) with different weights being used for different sizes of lettering. The size of the final output had to be considered in selecting the size of the lettering. Axis labels that were acceptable for an 8 x 10 inch chart were unreadable when the chart was reduced to a column size for inclusion in a journal. While the fonts initially provided had some flaws, they were close enough to the desired quality to be usable. More restraint was needed in the selection of shading patterns. This was especially important on bar charts where poor choices gave the illusion of curved or tilted bars. The shade patterns needed to be ordered from dark to light across the chart and the same angle (45 degrees) should be used for all patterns. Different patterns could be achieved by changing the thickness and spacing of the shade lines and by cross-hatching.

The ability to produce smooth curves was also needed and was no problem. In fact the illustrators were very impressed by the ability of the software to generate a smooth sine wave given only the peak and trough values. The windowing capability of the software was also greatly appreciated. This feature allowed the illustrators to reuse a portion of a chart, such as a company logo, on several charts. One feature that we were not able to provide was the ability to blank the interiors of curve marker symbols and the blanking of halos around curves and shaded markers. The illustrators decided they could wait for this feature as long as it would eventually be available.

Once the problem of meeting publication quality had been addressed, we looked at the other objectives for software. Machine- and device-independence were absolutely necessary since we were using different computers for development and production work and different devices for different applications. We had to support graphics terminals for previewing of plots, a COM recorder for color slides, and an electrostatic plotter for charts and posters. Having device-independent software saved a lot of time when new devices were added later, such as the color ink jet plotter which gave us the ability to produce color posters.

The fact that we were using different computers and different plotters made it more difficult to provide a user-friendly system. But because our illustrators were not computer professionals and should not be forced to be, it was necessary to make most of the operations transparent to the user. By using command files on our time-sharing computers, we were able to make the process used to access the graphics software appear the same regardless of which computer or plotter was to be used.

This system of command files made it easier to expand our software as needed. Because we knew our needs would change with time, we had to be careful to select software that was capable of growth. By also choosing software that was compatible with the existing computer graphics applications within UCC-ND, we were able to pass the quality enhancements we made for the illustrators on to the rest of the corporation. Sharing the software meant that the costs of development and maintenance were also shared. This meant that the graphic arts department did not have to bear the entire cost of the software as they would have had to do if a dedicated system had been chosen.

The objectives we identified were not met by a single piece of software but by a combination of software tied together into a system. There were several components in the system. The machine- and device-independent graphics subroutine library continued to be used for scientific and business applications and as a base for development of designer software. A building block front-end to the library was acquired from another national laboratory and enhanced to meet our publication-quality standards. Automatic prompting front-ends were developed for standardized applications such as bar charts and pie charts. These front-ends were suitable even for casual, infrequent users of the system. Automatic generation and submission of job control language for batch production runs were made part of the system. The final component of the system was the addition of menu-driven software for digitizing and graphic input.

While we did not totally meet all of our objectives, we did achieve a very satisfactory system as indicated by the usage of just one of the components. The building block front-end is accessed from three to six thousand times a month on one of our time-sharing computers. The graphic arts departments are also pleased with the statistics they have gathered. The most charts produced in one month by a single illustrator is 359. In the past year, the average units per person has gone from 80 to 180. The averages come from a group of illustrators with experience in the computer system varying from 5 years to 18 months.

4. HUMAN RELATIONS

With the major decisions relating to hardware and software taken care of, we should have had smooth sailing. We did not. We had ignored a critical area--human relations. Most of the problems we had were created by mistakes on the part of the programmers. After all, computer science is of itself a fairly useless endeavor. It is only when the computer is used as a tool for some application that its existence is justified. We initially failed to recognize that the application was not computer graphics but graphics and that the illustrators were the experts. This is a very common failing of programmers involved in computer graphics. We identified five causes of the problems we had in introducing computer graphics into the graphic arts department.

(1) Jargon differences created a barrier to communication. Computer science is full of violent sounding terminology. The illustrators (like anyone unfamiliar with computers) were somewhat wary of the system and afraid that dire consequences would ensue if they pushed the wrong button. So there they sat in the terminal room moving a cursor around, having just been warned that the operator would kill them if they were quiet too long. And to reassure them, we said, "The worst that can happen is that your job will bomb, but don't worry--after you crash and burn a few times, you'll get the hang of it." That was just the beginning. We realized we shouldn't use abbreviations (like .GE. for greater than or equal to) after one of the illustrators called to ask what "General Electric zero" meant. Still another term was discovered to be jargon after one of our programmers tried to order the "default soup" in a restaurant. Defaulting to preset values when arguments were skipped was one of the prominent features of our front-end software. On checking with the illustrators, we found that they had not been taking advantage of this feature. Reading the dictionary definition of "default" had caused them to interpret our description of a special feature as a warning of a potential problem. However, jargon was not the only communication problem we had.

(2) Unintentional insults were frequent because of each group's ignorance of the other. Again the programmers were the main offenders. In promoting the use of computer graphics to those outside of graphic arts (mainly to researchers and managers), we often stressed the superiority of computer graphics over visuals done "by hand". This argument was valid for the researchers and managers who usually had no artistic training. However, the same presentation was a slap in the face to the illustrators who did all of their work "by hand". A less serious offense was our referring to all members of the graphic arts department as "artists". We later discovered that the terms "artist", "illustrator" and "designer" had job level connotations and that we had not endeared ourselves to the senior designers of the department by introducing them as "graphic artists". Not all of our problems were caused by things we did; some were caused by things we didn't do.

(3) Ignoring the fears and prejudices of the illustrators caused those feelings to grow. The introduction of a computer to perform tasks formerly done solely by manual methods will cause concerns for job security in almost any profession and the graphic arts department was no exception. In addition to fears of losing their jobs, some members worried that they would lose the job level they had attained through years of hard work and training. There was also the fear that everyone would be "forced" to use computer graphics. This was especially strong among those who had chosen design as a career because they disliked mathematics and other like sciences.

(4) Overselling of computer graphics fueled these fears and caused resentment. W typical pride in the system we had implemented, we praised the software too highly, claiming that computer graphics could do anything, and that anyone could use computer graphics because it was so easy. We rode into the graphic arts dertment on a white horse, waving the banner of computer graphics and expecting to be welcomed with open arms. We were lucky they didn't ride us out on a rail because our claims were full of holes. Since our old software had been so much harder to use, the new system seemed easy by comparison. But the system was not all that easy. As one illustrator put it, "It's not easy; it's hard, but it's worth it." The claims that computer graphics could do anything arose from another cause, the last problem that we discovered.

(5) Computer scientists have not been trained to recognize presentation- and publication-quality graphics. Too often in computer graphics (especially that produced by programmers), the medium was the message. We had not learned moderation. We had not realized that for effective communication, the audience should not notice our slides but rather the information on the slides. We were like a child with his first box of paints--too enthralled in splashing colors around to worry what the final product looked like.

Identifying all these problems has been a learning experience for us. In retrospect we have identified necessary actions for establishing good relations in a project of this type. (This is fortunate for us since there are still other graphic arts departments within UCC-ND that have not yet added computer graphics to their services.) We eventually took all of the following actions, just not as soon as we should have.

- o We identified jargon so it could be explained or removed. We did this by consulting a technical editor unfamiliar with graphics or computer terminology.
- o We found someone in the graphic arts department with whom we could have an egoless discussion to determine our areas of ignorance.
- o We identified the fears and prejudices that existed and set up an open forum to discuss them. Management did much to relieve the fears, especially by pointing out that the graphic arts department was not converting to computer graphics, but merely adding it as a new service.
- o We realized the limitations of computer graphics. Because of the effort involved, some things just weren't worth doing. And because of software and hardware limitations, there were some things that could not be done.
- o We tapped the resources of the graphic arts department (and of photography and reproduction) to learn publication- and presentation-quality requirements.

5. CONCLUSION

Attention to these areas (although belated) has helped make our project a success. During the past four years, the number of illustrators using computer graphics full time has grown from one to eight, with more currently being trained. These illustrators are located at two different graphic arts departments with more departments to be added later. The final success of our project depended on careful attention to human relations as well as to the more obvious problem areas of hardware and software.

6. PUBLICATIONS

Ms. Riley wrote the following publications under her former name of Betsy Anne Clark.

"Automation and the Graphic Artist", presentsed December 1983 at 2nd National Conference on Computer Graphics: Delivering the Promised Productivity, held in Washington, DC; also published in proceedings.

"A Case Study: Respect Graphic Artists' Expertise When Implementing Computers", published March/April 1983 in Computer Graphics News, vol. 2, no. 6.

"Computer Graphics and Graphic Artists: A Rocky Courtship", 1982 NCGA proceedings.

"The Role of Human Factors in Introducing Graphic Artists to Computer Graphics", presented March 1983 at CAMP '83 (Computer Applications for Management and Productivity) held in West Berlin; also published in proceedings.

"Software-Hardware: Guides for Computer Graphics Use", published February 1983 in The Office, Magazine of Management-Equipment-Automation.

7. BIOGRAPHY

Ms. Riley has been involved in computer graphics since she joined the Computer Services organization at Oak Ridge National Laboratory ten years ago. Her section is currently responsible for support of computer graphics for the Department of Energy facilities in Oak Ridge. Support activities include maintenance, documentation, training, hardware and software consulting and general-purpose tool development. Ms. Riley pioneered the efforts to introduce the use of computers to graphic designers at Oak Ridge. She is a member of ACM, SIGGRAPH, and the National Computer Graphics Association.

20

GRAPHIC STANDARDS: STATUS & IMPACT

Dr. Barry J. Shepherd
IBM Corporation
11400 Burnet Road
Austin, Texas 78758 USA

ABSTRACT

There are many graphic related standards under development at this time. Their goals, status and schedules will be presented. The relationship between the standards and the topics of this conference will be explored. Sources for added information will be identified.

INTRODUCTION

This presentation has two major parts following this introduction. The first part lists the numerous graphic related standards now under development. The function and status of each standard (as of late 1984) is included. The second part of the paper discusses the relationship of graphic standards to several of the topics listed in the call for papers for this meeting. The views expressed will be mine, as an individual. However, I have been involved with graphic standard development since 1977 and am now Vice-Chairman of X3H3, the graphic standards committee in the USA.

GRAPHIC STANDARDS: STATUS

Nine different graphic related standards will be discussed below. Although my focus will be technical committees which operate under the rules of ANSI (the American National Standards Institute) there is close liaison with working groups within ISO (the International Standards Organization). There is another comment which applies to all of the standards to be discussed. All ANSI standards are expected to support (or use) existing ANSI standards. Thus, the goal is an integrated family of standards.

GKS

GKS (the Graphic Kernel System) is now an ANSI and ISO standard. It defines functionally at the application programming interface. It is device independent, host system independent and application independent. It currently supports 2 dimensional output and input, but a fully compatible extension for 3 dimensional data (and viewing) is expected to be approved in 1985. Other extensions may be considered by X3H35, a subgroup of X3H3. The ANSI version does have an additional smaller level of output capability than the three levels provided by ISO GKS.

PHIGS

PHIGS is the programmers hierarchical interactive graphic system. The significant extensions it offers beyond GKS-3D are in its hierarchical segmentation support and the dynamic ability to modify segment contents and relationships. In general, standards codify existing practice rather than define new technology. PHIGS, however, is intended to support today's high function workstations. It is thus well matched to CAD/CAM applications, although it does not contain any CAD/CAM specific functions. A revised draft may be cleared by X3H3 for public review late in 1985. This work is being done by X3H31, a subgroup of X3H3.

LANGUAGE BINDINGS

The two standards above are not only device independent, they are also programming language independent. They are a functional specification of what an application programmer can do. How a specific capability is requested in a given programming language is specified by a language binding. A binding specifies function names and data types. In most cases, the parameters lists are taken from GKS. However, some modern languages such as Ada, Pascal, and C provide more complex data structures than available in Fortran and this is reflected in the bindings.

As Chairman of the ISO GKS language binding effort, I will continue to keep the ISO and ANSI bindings as similar as possible. We have a final common Fortran binding. The Pascal binding is now out on public review and should be approved during 1985. The Ada binding should go out for public comment in early 1985. Although there is not yet a C language standard, there should be agreement on a C language binding early in 1985. A revised binding to PL/I should be available late in 1985.

The language binding work in the USA is being led by X3H34, another subgroup of X3H3. All of the bindings to date employ a "call-like" interface. There will also be bindings developed for GKS-3D, PHIGS and VDI (the virtual device interface). In time, the various languages may extend their syntax and data types to provide "native language" support of graphics.

VDM

The Virtual Device Metafile (VDM) defines the functions needed to describe a picture. The description can be stored or transmitted to some other location. The VDM supports output only and easily allows capture of GKS pictures. This is an example of one standard supporting another. This should not be surprising, since VDM was developed by X3H33, a subgroup of X3H3. It should also be no surprise that the VDM is a functional specification.

There are three bindings of VDM which are standardized. They are all data stream (not programming language) oriented and use ASCII. It should be noted that the VDM contains no segmentation (i.e., it is one smooth piece, without lumps). This means that the VDM is not the way to store a graphic picture for later revision. The developers of PHIGS are studying an archive (save/restore) capability which could support revisable picture storage within a single system.

The VDM went out on public review early in 1984. As a result of comments, it was extensively revised and functionally enhanced. It will complete its second public review early in 1985. It is now called the computer graphics metafile (CGM).

VDI

The virtual device interface (VDI) lies between GKS (and hopefully PHIGS, in time) and the devices in a graphic system. While GKS is device independent, devices are not all the same. The VDI is the lowest device independent interface in a graphic system. It shares many characteristics with the CGM, since it too is being developed by X3H33. It is a functional specification but will have both procedural bindings (like GKS and PHIGS) and data stream bindings (like CGM).

The VDI is a low level interface. It is well matched to dumb displays (or plotters) and GKS. It is not well matched to intelligent workstations, at the level of the conceptual GKS workstation. It is also not well matched to a distributed or network environment. The VDI should be available for public review in 1985. It has been renamed the computer graphics interface (CGI).

IGES

The Initial Graphic Exchange Specification is not viewed as a graphic standard by X3H3. It is obviously capable of transmitting graphic pictures, like the VDM. However, IGES does not have multiple bindings. It does have a specific application (or application class) focus. IGES also provides for transmission of graphic product definition data which is outside the scope of the VDM.

NAPLPS

The North American Presentation Level Protocol Syntax is another standard which X3H3 views as not supporting general purpose graphics. It is an interface with a single binding which is optimized for data (text and graphic) presentation, on one class of device. Its very limited graphic input capabilities prevent its use as a general graphics interface.

OFFICE SYSTEMS/TEXT PROCESSING

There are two standards activities which are working to integrate graphics into text documents. Within ISO, a recent reorganization will combine the work on Office Systems with that of Computer Languages for processing text. ANSC X3H3 will continue the liaison activity it has already established. We would expect the CGM (or possibly GKS) to provide the graphic capability desired by these other committees (specifically ANSC X3V1). We would be concerned if committees outside X3H3 tried to define graphic capability without our assistance. Drafts of the graphic portion of these non X3H3 standards may not be completed until 1986.

OPEN SYSTEMS INTERCONNECTION

OSI is a seven layer model for communication between two possibly different intelligent processes. The virtual terminal is a definition of one of the top interfaces, when the receiving process is a terminal. This work has occurred primarily in the international (ISO) arena. The work on a graphics virtual terminal specification is just beginning. The ISO reorganization mentioned above has brought the virtual terminal and graphic standard work into the same organization. I hope this will lead to a consistent graphic virtual terminal specification.

GRAPHIC STANDARDS: IMPACT

In this section, I will relate the standards listed above to some of the topics stated to be of interest to the attendees at this conference. I will not discuss optical disks or mass storage systems, since they are not affected by graphic standards. However, there is relevant standardization activity in these areas.

PRODUCTIVITY ADVANCEMENT

Both GKS and PHIGS enhance the productivity of application developers by providing a consistent vocabulary and conceptual model for graphic systems. The language bindings are matched to the capabilities and style of the language they support. Thus, whatever language the application developer uses, the graphics capability will be a natural adjunct.

RASTER TO VECTOR CONVERSION

Both GKS and PHIGS provide a cell array output primitive, which generalizes raster output capability. Raster input is absent from both these standards, as well as from the CGI. However, the CGM could be used to provide raster input to a raster to vector conversion program. The CGM could also be used to output the vector description, although the PHIGS archive file might be a better candidate, if graphic editing was a subsequent goal.

IDENTIFYING FUTURE REQUIREMENTS

In this area, standards can rarely provide input. That is because most standards codify existing practice, rather than define new technology. However, requirements from groups such as the attendees at this meeting can identify standards which need to be developed. I welcome your ideas on new/future graphic standards.

PRODUCTION AND DISPLAY OF ENGINEERING DATA

Applications built on GKS, PHIGS, and CGI will support the creation, revision and display of graphical engineering data. The CGM supports the storage, retrieval and distribution of the same data.

DISTRIBUTED CAD/CAM

The CGM and CGI are adequate to support the presentation of engineering drawings in a distributed environment. To adequately utilize advanced function workstations, a high level CGI, with more function than is proposed in the current standard, must be defined. If distributed development and revision of drawings is desired, then PHIGS should be the application base. As mentioned previously, CGM does not allow revision and thus is an inhibiter to this strategy.

CAD IN VARIOUS APPLICATIONS

CAD and other applications can be built on GKS, PHIGS or even the CGI. However, all of these standards are application independent. Is there a need for additional application specific extensions to these standards? Is there a body of common practice from which to evolve a standard? Would the economic benefits to users and vendors exceed the development costs? Who would develop, maintain, interpret and revise the extensions? I welcome your input on these questions.

TECHNICAL DOCUMENTATION

Technical documentation including text and graphics should be supported by standards in the office systems area. I would expect the graphic portion of text and graphics documents to come from applications based on GKS, PHIGS or the CGI. As indicated above, a revisable CGM supporting PHIGS is needed.

CONCLUSIONS

Some graphic standards have been adopted. Other standards are nearing completion. These standards form the basis of a family of related graphic standards. Their impact on applications which support the creation, revision and distribution of graphic data will be major. In the case of Ada based systems it will be unavoidable. However, additional standards are required if graphics is to reach its full potential. The most pressing need is for a revisable CGM and a high level CGI.

REFERENCES

There is an ever growing set of references on GKS and other graphic standards. There is a very readable introduction to GKS by Hopgood, Duce, Gallop, and Sutcliff. There is a more rigorous treatment by Enderle, Kansy, and Pfaff. The GKS and CGM standards are available from ANSI headquarters in New York. Information packages on the CGI, PHIGS, and language bindings are available from CBEMA in Washington, D.C. There are tutorials at both NCGA and SIGGRAPH annual conferences. Finally, there are articles in the trade press such as Computer Graphics World, IEEE Computer Graphics and Applications, Mini Micro Systems, Electronic Design, and Byte. In general, you should not trust the accuracy of articles whose author(s) are not on the graphic standard committee they are describing.

BIOGRAPHY

Dr. Shepherd has been active in the architecture, design, and implementation of graphic systems for more than 15 years. Currently a senior programmer with the Entry Systems Division at IBM Austin, he has numerous publications to his credit. He has long been interested in graphic standards, and is now vice chairman of the ANSI graphic standards technical committee, X3H3. He is also the leader of the ISO language binding activities for GKS. A founding vice president of NCGA, he is a senior member of IEEE and a member of ACM and SIGGRAPH.

21

GRAPHIC RELATED STANDARDS

Dr. Barry J. Shepherd
IBM Corporation
11400 Burnet Road
Austin, Texas 78758 USA

Abstract

This paper will provide an overview of the many standards under development which relate to computer graphics. This paper will not restrict itself to the work of ANSI Technical Committee X3H3 (graphic languages), although the committee's activities will be reported. Most of the countries in Europe have active graphic standards projects at both the national and international levels. The relationships between the various European activities is similar to that described below. The pervasive nature of computer graphics and its growing importance have caused many other standards committees to also consider graphics. The opinions expressed in this paper are those of the author, recently appointed vice chairman of ANSI X3H3. They do not necessarily reflect the views and policies of the standards committees referenced, nor of IBM.

1. STANDARDS CHARACTERISTICS AND DEVELOPMENT

In the United States, standards are developed jointly by representatives from industry and the affected user communities. Representatives from Government participate as equals during development; compliance with ANSI (American National Standards Institute) standards is voluntary. Standards codify existing practice rather than explore the frontiers of technology.

There are three cornerstones to the formal standards making process:

- o Participation is voluntary and open to all qualified individuals.
- o Consensus must be achieved on all matters (one vote per organization).
- o Public review and resolution of all issues are required before adoption.

The guidelines above encourage the development of fair standards. However, a widely accepted standard has several additional characteristics. These include:

- o Satisfying a real need (i.e., many suppliers and customers for the "product").
- o Implementable without breaking the laws of nature or man.
- o Allow innovation by not constraining the internal structure of implementations (standards define interfaces, not internals).
- o Provide economic benefits to affected suppliers and customers.
- o Allow exceptions (noncompliance) where appropriate.

2. A SIMPLE GRAPHIC SYSTEM

Figure 1 shows the main components of a graphic system and some interfaces which might be standardized. There are a few assumptions, common to modern graphic systems, which are reflected in the illustration. First, the management and manipulation of the "graphic model," which includes the data from which a display image is defined, is isolated from the graphic support package. This separation of modeling and display functions was an important insight of the mid-1970's graphic advances.

The second assumption is that the Application Programming Interface (API) is device independent. This means an application can produce a picture on a raster display, a directed beam display, an electrostatic plotter, or a pen plotter without any changes to its requests for graphics services. This is an example of information hiding, although an application can choose to optimize its picture description for a specific device, giving up device independence. The concept of device independence has been extended to graphic input, with six types of logical input device being defined.

3. ANSI GRAPHIC STANDARDS (X3H3)

After a study group verified the need for, and benefits of, potential graphic standards and existence of applicable technology, the American National Standards Institute (ANSI) established committee X3H3 to work on graphic programming languages. Since its formation in 1979, the committee has been chaired by Dr. Peter Bono. It now consists of representvatives from more than sixty different organizations who regularly attend meetings.

There are three subtask groups in ANSC X3H3 which are developing draft proposed American National Standards (dp ANSs). The formal development methodology adopted by X3H3 indicates why standards creation is time-consuming.

First, the subtask group writes a document listing technical issues, possible resolutions for each issue, pro and con arguments for each issue, and the recommended resolution for each issue. Once all of X3H3 agrees on the issue resolution, the subtask group writes a draft standard. The draft is sent for review by X3H3, then X3, then the general public. At each stage, comments must be considered and either incorporated or reasons provided for their rejection.

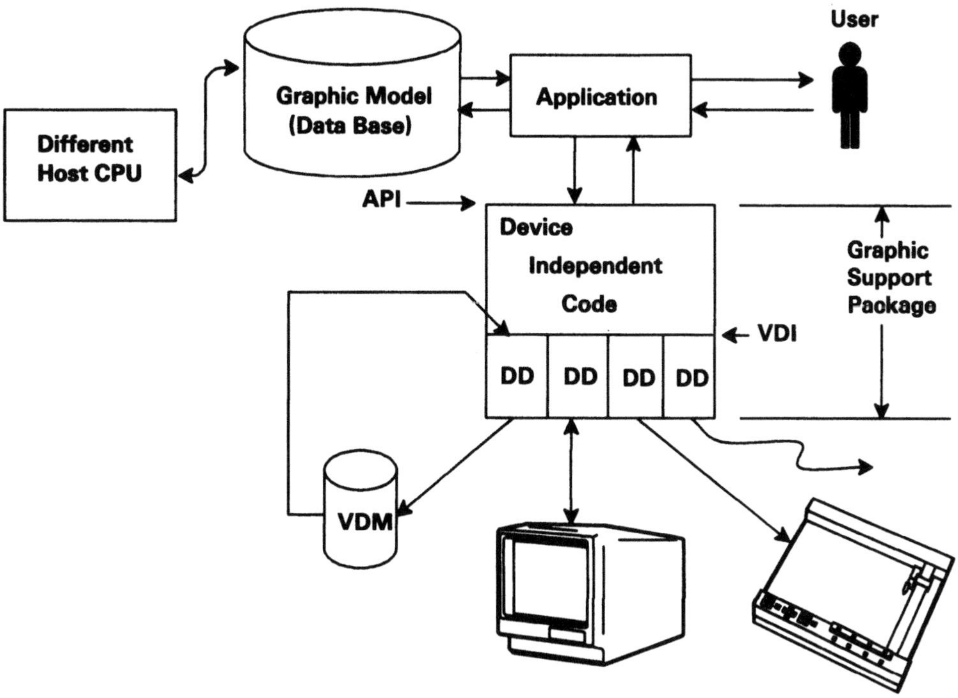

Figure 1--Graphic System Model

4. PROGRAMMERS' INTERFACE

This will be the most important of the several graphic standards that will be described. It is identified as the API in Figure 1. The first formal standard in this area is known as the Graphical Kernal System (GKS). This standard was develop0ed within TC97 SC5 WG2 of the International Standards Organization (ISO). The technical work was led by the German Standards Institute (DIN). Their efforts were influenced by two landmark documents (not formal standards) developed by the Graphic Standards Planning Committee (GSPC) of the Association for Computing Machinery Special Interest Group on graphics. The two ACM SigGraph CORE reports were published in 1977 and 1979.

The standards wheel has now gone full circle witlh the decision by ANSC X3H3 to formally process the ISO Standard (GKS) as a United States standard. There is one change to GKS whilch is included in the ANSI version: A new level of output capability smaller than any of the ISO GKS levels. This new level is a proper subset of the lowest ISO GKS level. Thus, in the U.S., there are four output levels instead of three. The three levels of graphic input capability are the same worldwide, with "no input" as one of the levels.

GKS includes the following characteristics:

- o Device-independent programmers' interface.
- o Only two-dimensional primitives and transformations.
- o Elegant, device-independent graphic input (optional).
- o There is no concept of current position, and thus no "relative" primitives (unlike the CORE proposals).
- o No application-specific functions are provided.
- o Unique device capabilities can be accessed via escape and generalized drawing primitive (GDP).
- o The functionality is not bound to any specific language.

The last point in the list means that additional efforts are required to produce a standard FORTRAN graphic interface, a standard PASCAL graphic interface, and so on. However, all of these language-binding efforts will start from a common statement of the functionality to be supported--GKS. For some languages, standard graphics may be added (bound) in the form of subroutine calls. For other languages, graphics may be supported by means of extensions to the existing syntax on an equal basis witlh the original language.

5. BASIC

The new extended BASIC programming language standard presents an interesting contrast to the above discussion. This standard, currently under revision following public review, provides simple graphic output and input. However, the functionality for graphics was defined before X3H3 had finished their document. It should be no surprise to learn that the two proposals do not coincide. After considerable work by both groups, there were few differences. They included:

- o BASIC provided no marker primitive.
- o BASIC supported current position.
- o BASIC supported only one device without selection from an application and, thus, was not device independent.

The graphic capabilities in BASIC had not only the syntax of BASIC but also some of its semantics, such as the effect of a semicolon. This was a benefit (requirement) from the BASIC standards group viewpoint and an odd side effect for graphic purists. One motivation for the definition of the new GKS subset level by ANSI was to standardize a set of functions small enough to satisfy BASIC and still be consistent witlh modern graphic concepts.

The international standards organization has now required BASIC to incorporate graphics capability which is directly drawn from GKS.

6. PROGRAMMERS' HIERARCHICAL INTERACTIVE GRAPHICS STANDARD

Within ANSC X3H3 there is a subgroup working on a new application programming interface, called PHIGs. The functionality of this interface is greater than that of the highest GKS level. It includes three-dimensional output primitives and transformations. It also includes a richer set of segmentation capacilities than that provided by GKS. Part of this richness is greater (dynamic) control over the visual appearance attributes (line style, color, etc.) of primitives within a segment.

The goal is to support the most capable intelligent displays which are starting to become available. The target hardware will have capabilities such as hidden surface removal, three-dimensional transofrmations, and linear shading of surface patches--all in realtime. The software will support the interactive use of such advanced display hardware.

7. VIRTUAL DEVICE INTERFACE

The virtual device interface, indicated as the VDI in Figure 1, is an interface which is internal to a graphic support package. As described above, the API is device independent, but many devices have different characteristics. The VDI is the last point (closest to physcial devices) where device independence can still be maintained. This interface is being developed by a subtask group of ASNI X3H3. Their work includes the following concepts:

- o Device independence for output and input.
- o Partitioned into required and nonrequired sets of function.
- o Data interface, not a procedural interface.
- o Some orders have variable length data fields.
- o Consistent witlh the Open Systems Interconnection (OSI) model.

As with the API proposals, the VDI definition will not be specific to any protocol. However, as an existence proof of feasibility, one binding will be exhibited. Because new ANSI standards must align with existing standards, the sample binding must be based on the ASCII character set. The current suggestion is to start from the NAPLPS (videotex) proposal described below. Just as there will be multiple bindings of the API functions to various programming languages, there will also be multiple bindings of the VDI to different protocols. In fact, some devices may accept and directly execute VDI orders.

8. VIRTUAL DEVICE METAFILE (VDM)

The same subtask group of ANSC X3H3 which is working on the VDI definition also developed the VDM. The virtual device metafile is a way to transport the definition of a picture over space and/or time. The VDM will consist of all the required set of VDI functions and a fixed selection of the non-required VDI functions.

There are two obvious differences between the VDM and the VDI. First, the VDM supports only output functions. Second, because the creation and use (interpretation) of the VDM may be separated in space and/or time, the error handling philosophy must differ from that of the VDI. One point of similarity is the multiple bindings of the VDM functionality included in the draft standard now on public review.

9. NORTH AMERICAN PRESENTATION LEVEL PROTOCOL SYNTAX

We refer to this unwieldy named standard as either NAPLPS or simply videotex. Although it is at the same functional level as the VDI in Figure 1, videotex differs from the work of X3H3 in several important ways. First, it is application specific to videotex, an information distribution system which employs raster displays, such as TV sets. Second, although graphic capability is part of the videotex repertoire, only the most rudimentary input capability is provided (unlike the graphic VDI discussed earlier). Finally, it is a data stream definition rather than a programming (software) interface.

The videotex proposal has been standardized by ANSI X3L2, which deals with character set codes. The work relies heavily on the Picture Description Instructions (PDI) definition produced by the Canadian Telecommunication Research Center. The NAPLPS differs in some details from a European videotex proposal which is known by the name of its sponsor, CEPT.

10. APT

The language of Automatic Programmed Tools is a mature ANSI standard which obviously addresses a specific application area. Although it may be stretching matters, the parts produced by a computer-controlled milling machine are an obvious example of three-dimensional output. (Incidentally, the output of APT is a device-independent file that is then made device-specific by a postprocessor. This is the same structure that is presented in Figure 1, with the postporcessors corresponding to device drivers.)

There are new ways in which graphics is coming to APT. One use is to display the actions of a machine tool on a grpahic display. To accomplish this, a postprocessor converts the APT output into graphic commands at either the API or VDI shown in Figure 1. Another use is an interactive graphic application which allows incremental construction of the definition of the object to be machined. The interactive application would presumably be most easily written using the graphic API, but its output would be an APT output file, not the grpahic VDM.

11. IGES

The Initial Graphic Exchange Specification is unusual in that it was a standard sponsored by the National Bureau of Standards. Normally the NBS, despite its name, administers, maintains, and distributes standards rather than developing them. The work has now been turned into an official ANSI standard by committee Y14.26.

The subject matter of IGES is specifically the interchange of mechanical design and manufacturing (product spescification) data. This is indicated in Figure 1 as the connection between the graphic model and a different host CPU. IGES is specific to mechanical design because there is no universal graphic model. It must be realized that the primitives which IGES uses to describe objects are not the same as those used by solid modeling systems (which are not yet being standardized).

There may be some apparent similarity between the capabilities provided by IGES and those provided by the VDI. However, that similarity is overshadowed by the differences. One major difference is that IGES does not provide any input functionality--it is unidirectional. Also, IGES is a classic example of the process-to-process communication described by the OSI model. Finally, IGES primitives cannot be directly displayed graphically, let alone be edited (in the sense that a vector graphic image can be edited).

12. COMPUTER LANGUAGES FOR PROCESSING TEXT

Like APT, IGES, and NAPLPS, the CLPT (Computer Languages for Processing Text) standards activities are application specific. ANSI committee X3J6 addresses standards for the entry, editing, and formatting of text documents. They propose the use of descriptive document markup ("this is a paragraph") rather than a procedural markup ("indent 10 spaces"). Thus, much of their work relates to the model (data base) structure used to contain text documents and the different objects which it may contain.

Graphics relate to their activities in potentially three areas. First, in common noncomputer usage, text documents include line art. This line art could be included as objects described in terms of graphic VDM prmitives. A second potential area is the use of the graphic VDM as the interface between formatting programs and physical output devices. This proposal comes from the belief of graphics people that everything (including justified variable spaced multifont text) can be realized by means of graphic primitives. Finally, use of the graphic VDM/VDI would automatically provide soft-copy proofing on graphic displays.

13. OFFICE SYSTEMS

Especially within ISO, work has expanded recently in ISO TC97 SC18 on a set of related standards dealing with various aspects of office systems, including the preparation, distribution, and presentation of text documents. One issue to be resolved is the apparent overlap with the work of the CLPT group. However, the office system group further claims responsibility for all text documents, which it specifically defines to include vector graphics and even image. The claim of responsibility over graphics will have to be resolved with ISO TC97 SC5 WG2 which feels it owns graphics (which includes text). It is too early to predict the outcome of these jurisdictional disputes, but the importance of graphics in the application area called office systems is verified by the hardware and software already available.

14. FACSIMILE

Especially in Europe and Asia, facsimile standards and devices are of ongoing commercial interest. If one defines facsimile as a raster image (normally produced by a scanner), then office systems and graphic standards both incorporate a generic facsimile capability, at least for output.

In fact some people have noted that most new technologies for both displays and printers are based on raster concepts. Thus, raster image could be considered as the data stream used to access all output devices. One problem with this proposal is the size of raster image data streams, compared to the vector and alphanumeric data which consititute its true information content. This is especially serious for devices which are not locally (high bandwidth) attached to their supporting host system. Second, image data is almost impossible to animate, or even edit, without4 expending an inordinate computational effort.

15. CONCERNS

From the above descriptions, one can infer that there is a lot of ongoing standardization relating to graphics. Also, because new standards activities are proposed "from below" by technical experts, rather than being requested "from above" by ANSI Headquarters, it is very difficult to provide adequate coordination among the many efforts (i.e., no published architecture which demonstrates their interrelationships). The result is not only overlap among various efforts, but also leaves areas not yet addressed where standards might expect to be required, such as:

- o Application specific support functions (API extensions) for CAD,
- o API extensions for business graphics, and
- o Generic data base model structure for business graphs.

For those standards already defined, there are the related questions of conformance and certification. Regarding conformance of programming language standards, the concept of conforming applications and conforming implementations has arisen. A conforming application uses, as a maximum, the function included in some (level of a) standard. A conforming implementation provides, as a minimum, the function included in some (level of a) standard. A conforming application can be transported with minimal effort between conforming implementations. This concept is considered unacceptable by ISO, which demands implementations provide neither enhancements nor omissions relative to the standard.

Assuming agreement on the meaning of conformance can be achieved, the question of certification of implementations must be answered. Because of a recent U.S. Supreme Court decision holding standards committees liable for "incorrect" interpretations (ASME boiler safety device), standards-making bodies in the U.S.A. have withdrawn from commenting on the adequacy of implementations. Unless some user-based organization steps forward to assume the task, customers will have only the reputation of an implementor to reassure them about the fidelity of an implementation to standard. This is not a problem in Europe where, with government support, certification procedures are a subject of considerable current interest.

16. CONCLUSIONS

Some excellent work has been done in the area of graphic standards. Based on current experience, even more technical committees will begin graphic-related standards efforts. Increased cooperation among the committees must be encouraged if the user community is to receive a coherent set of standards. The fact that some standards are nearly completed and others barely started does not ease the problem.

17. REFERENCES

There is an ever-growing set of references on GKS and other graphic standards. There is a very readable introduction to GKS by Hopgood, Duce, Gallop, Sutcliff. There is a more rigorous book by Enderle. The GKS and VDM standards are available from ANSI in New York. There are also tutorials at both the NCGA and ACM SigGraph annual conferences. Finally, there are articles in the trade press, such as Computer Graphics World, IEEE Computer Graphics and Applications, Mini Micro Systems, Electronic Design, and Byte.

18. BIOGRAPHY

Dr. Shepherd has been active in the architecture, design, and implementation of graphic systems for more than 15 years. His doctorate is in nuclear physics, from the University of Washington. He is now a senior programmer with IBM's Entry Systems Division in Austin, Tx. He has long been interested in graphic standards and is now vice chairman of the ANSI TC on graphic standards (X3H3) as well as heading up the ISO GKS language binding work. A founding vice president of NCGA, he is also a senior member of IEEE and a member of ACM and SigGraph.

22

CONSIDERATIONS FOR ESTABLISHING A DISTRIBUTED COMPUTER GRAPHICS MAPPING SYSTEM

Larry D. Sieck, CEO
Cochise Incorporated
La Mesa, California 92041 U.S.A.

1. INTRODUCTION

The establishment of a distributed computer graphics mapping system has been the goal of numerous organizations since the early 1970's. The planned benefits from such a system have included eliminating redundant mapping, improving map accuracy and timeliness, providing a single source (database) for map information, lowering costs and increasing map producer/user productivity. Although the benefits to be realized today are still very much the same, the distributed mapping systems have evolved from simple drafting functions to some now approaching full expert system capabilities.

2. DISTRIBUTED MAPPING SYSTEMS

In the early 1970's a distributed mapping system consisted of a computer with workstations remotely attached using a telecommunications line. Initially map data was transmitted at 4800 bits per second (BPS). Response times were very slow at this speed and user complaints resulted in data being transmitted at 9600 BPS. This speed was still much slower than that in workstations attached directly to the computer. In the late 70's dedicated lines were used which approximated the speed of a direct tie to the computer. Response times still varied considerably, depending on the functions being executed at each workstation and the number of workstations tied to the computer. The next major step was the introduction of minicomputers at each workstation to handle local graphic manipulations. These workstations were referred to as "intelligent workstations." Workstations both intelligent and dumb, that are tied either directly or via a telecommunications line to a computer, are said to be configured in a "star" network. A star network is shown in Figure 1.

With the star network, workstations that were geographically dispersed could, via telecommunication lines, access the same common database. Additional intelligent workstations could be added to the star network without appreciably degrading response time. One main disadvantage of the star network is that, with failure of the central computer, the whole network fails.

In the early 1980's standalone workstations were introduced. These have their own dedicated computer, memory, disk storage and displays. They can be operated individually or in a "ring" network as shown in Figure 2.

One advantage of the ring network is that map data can be passed around the ring to other workstations. Another advantage is that if one of the workstations is down, the rest remain operational. A disadvantage is that there is no central database. If several workstations contain the same map data, each has to be updated simultaneously or the map data will shift out of synch. And ring networks are restricted to a local area.

Although there are other network configurations available, the star and ring seem to be the most popular for mapping systems now.

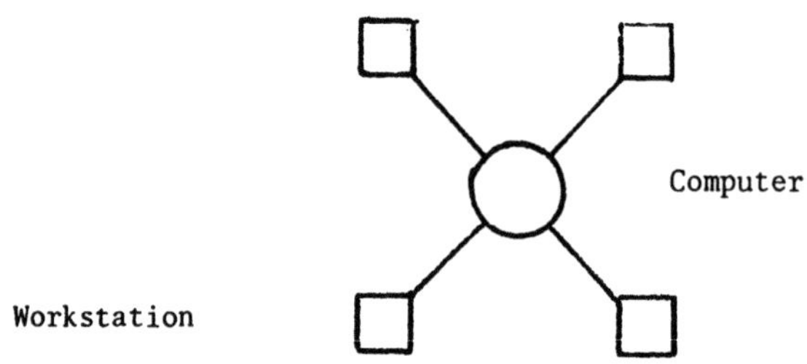

Figure 1 STAR NETWORK

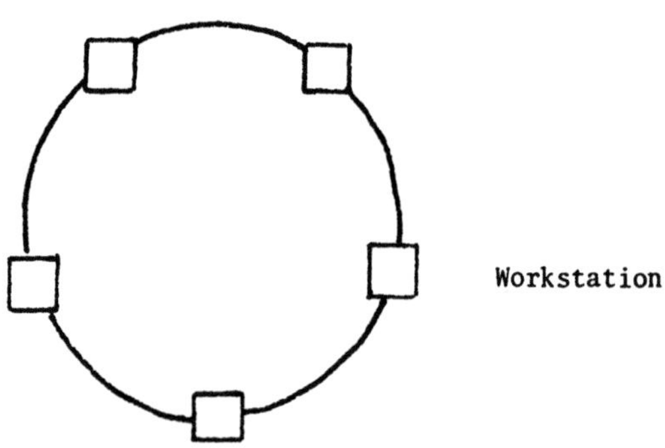

Figure 2 RING NETWORK

3. DISTRIBUTED MAPPING CONSIDERATIONS

The decision to establish a distributed computer graphic mapping system requires that several factors be addressed. Not all of the factors will necessarily impact the network configuration, but all of the factors must be addressed if the mapping system is to be successful. The factors are:

1. Map Products--The source for all map information must be defined. All analysis programs and reports that will utilize the map database must be identified along with computing, core memory, data storage, and frequency of use requirements. The map products have to be created, then approved by the users, and update frequencies scheduled, as well.

   Results of the above will help to determine where map conversion and updating should take place, and knowledge of the analysis program and report requirements will help to establish computing, core memory and data storage requirements. Identification of the map products and update requirements will indicate where output devices should be located.

2. Map Database Integrity--Map data security, data integrity and privacy requirements must be protected. For example, those who will be allowed to access, change or update the map database need to be identified, and appropriate passwords should be established for read, write, create, delete and erase functions.

3. Other Applications--An organization's plan for office automation should be carefully determined. Most organizations already employ word processing, electronic mail, computer generated reports or charts, electronic calendars, and so on. Often these are supported by separate systems. The addition of a computer graphics mapping system should not entail installation of equipment and functions that duplicate or fail to communicate with what is already in place.

   One large organization analyzed its requirements for systems to support word processing, personnel management, non-graphic analysis, and computer mapping. In one suggested plan, independent systems were to be installed to support each of these operations with the result that many employees were to receive 4 terminals, each using its own communication network. In fact, upon review of such proposals, the organization discovered that several employees already had two terminals assigned to their use. It is clear that in redundant or incompatible systems, efficiency is significantly reduced.

4. Performance--When designing a system, both mapping and non-mapping computer requirements need to be addressed. The consequence of failing to do this is a varying performance

in the installed system. It is frustrating for a user to execute a function, get a response within seconds one time, but wait for half a minute for the same function to be executed the second time. The user of a mapping system should have performance that is consistent and reliable. A user will accept the fact that some functions take longer to execute than others, but if the execution time for the same function varies drastically, the user may become discouraged and lose interest in the system.

5. Data Exchange--Within the same organization, or among organizations, there is often the need to exchange map data created on different systems. One method for data exchange is to obtain a translator program, should it exist, from the system vendor. In many cases, it is necessary for the user to write his own translation program or contract to have this done.

What is needed is a standard that can fully handle the exchange of map data created on different systems. The Initial Graphic Exchange Standard (IGES) is supported by most major system vendors. IGES is very good for translating graphic data, but it currently does not support the full exchange of non-graphic attribute data. Several mapping organizations are encouraging the IGES committee to include both graphic and non-graphic mapping data. One mapping organization that has been actively pursuing a mapping standard for IGES is the Southern California Computer Aided Mapping Association.

4. CONCLUSION

This paper has addressed some of the major factors that should be considered when establishing a distributed computer graphics mapping system. The two communications networks commonly used in mapping systems, star and ring configuations, can be used independently or may be combined. The choice of which network to use must be considered in light of the organization's total needs. These include response requirements, amount of data to be transmitted, user requirements and cost factors.

5. BIOGRAPHY

Mr. Sieck is the Chief Executive Officer of Cochise, Inc. and has been actively engaged in computer graphics for more than fifteen years. During the past five years he has been a computer graphics consultant, assisting the federal, public and private sectors.

Mr. Sieck is a lecturer and instructor on interactive computer graphic systems, currently lecturing at U.C. San Diego. He has a Bachelor of Business Administration degree from National University where he graduated Summa Cum Laude.

23

THE MARRIAGE BETWEEN CAD/CAM SYSTEMS AND ROBOTICS

Richard L. Simon
Director of CAM Product Management
Computervision Corp., Bedford, Mass., USA

1. INTRODUCTION

Manufacturing technology has been an evolution from the time of its birth, during the Industrial Revolution of the 1700's, until today. In this sense, the evolution of manufacturing technology is analogous to the "Theory of Evolution."

Each manufacturing technology or species is constantly improving and adapting to the demands of the changing environment. Technological improvements result in higher levels of manufacturing productivity and the ability to accomplish altogether new tasks. New manufacturing species are developed that involve new ideas, combinations of new and old ideas, or perhaps the synergy achieved from two old ideas. This basic pattern is also present with the technologies of CAD/CAM (Computer-Aided Design and Computer-Aided Manufacturing) and Robotics.

CAD/CAM and Robotics are two new species of manufacturing technology that have evolved over only the past twenty years. CAD/CAM saw its birth in the research labs of universities and in the automotive and aerospace industries during the 1960's. Because of the great productivity improvements of CAD/CAM on engineering and manufacturing. CAD/CAM has been called the "Third Industrial Revolution." Robotics, another productivity booster, also saw its beginnings in the early 1960's with the first major applications in the automotive industry.

CAD/AM and Robotics are still evolving in their own separate ways. Since both are now computer and software intensive, as well as young technologies, many improvements, encompassing new applications and higher levels of productivity, are yet to be seen. However, these separate paths have crossed. A new chapter in the evolution of manufacturing productivity improvement will result from the marriage of CAD/CAM and Robotics. This union will involve CAD/CAM robotic workplace design, graphic robotic programming and robotic simulation.

2. THE CHANGING ROBOTICS INDUSTRY -- TRENDS

The improvement in world productivity and product quality in the next decade will largely depend on new manufacturing technologies. Among emerging manufacturing automation technologies, Robotics is considered one of the best bets for achieving these objectives. Robotics will also reduce costs and allow many companies to remain competitive in their industries.

The Robot Institute of America defines a robot as:
"A reprogrammable, multifunction manipulator designed to move material, parts, tools, or specialized devices through variable programmed motions for the performance of a variety of tasks."

This definition can be expanded to include the control and syschronization of the equipment which the robot works with. This capability can eliminate the need for humans to work in the immediate environment, which may be hazardous.

Robotics fills the gap between special purpose automation and manual manufacturing. It is important to recognize that Robotics differs greatly from manual and special purpose automation, since a robot is capable of being programmed to automatically perform different tasks. Robotics, then, is a form of programmable automation.

Three factors have been shaping the future of the robotics industry. These include evolving robotics technology, broadening applications, and industrial acceptance.

The most significant trends in robotics technology today involves the computerization of controls, and advances in robotics software and sensory capabilities. Because of these trends, the dumb pick and place robots of the 1960's and 1970's have evolved into a smarter breed of highly programmable sensory robots of 1980's. The mechanical element of the robotic system, the manipulator, will be less dramatically affected.

Robotic applications have also been expanding. The limited capability pick and place robots of the 1960's opened up applications in hazardous environments. Examples include: automotive spot welding, simple material handling such as moving hot billets into and out of forging dies, paint spraying, etc. In these applications, sophisticated sensory capabilities and complex robotic programming were not necessary. Also, production volumes were large, justifying lengthy set-up times. Today, the evolving technologies mentioned above are permitting Robotics to be applied in the more diversified realm of small and medium batch manufacturing. Batch manufacturing represents the majority of world manufacturing. Robotics' move into batch manufacturing is technologically possible due to the advances in robot controllers, sensory capabilities and programming. New applications include seam welding, complex material handling and assembly.

The third factor shaping robotics for the future is industrial acceptance. Robotics is now recognized as an automated manufacturing tool for a wide variety of batch manufacturing applications. Labor and management understand that the competitive world environment demands that their companies implement the most productive methods of manufacturing, just in order to survive. Acceptance has not been consistent with the distribution of worldwide industrial production; however, a recent study showed that robots other than manual manipulator and fixed sequence types were distributed in the following manner.

Japan	- 58%	Rest of Western Europe	- 8%
USA	- 17%	Rest of World	-12%
West Germany	- 5%		

3. FROM CAD/CAM TO CIM INCLUDING ROBOTICS

The manufacturing side of CAD/CAM, CAM, has traditionally been applied to NC programming. As we move into the 1980's, the role of CAM in a CAD/CAM environment takes on increased importance, especially considering its position as an integrator in helping firms achieve the greater benefits of Computer Integrated Manufacturing (CIM).

The CIM concept encompases many manufacturing computer-based automation applications. CIM can be thought of as a closed-loop feedback system whose prime inputs are product requirements and whose prime outputs are finished products. CIM comprises a combination of software and hardware for product design, for production planning/control and for production processes.

CAD/CAM is an excellent CIM integrator for computer based applications in manufacturing. CAD/CAM's integration ability rests on a strong foundation that is the common engineering and manufacturing database. This allows engineering to define a product model. Manufacturing can use the same model for a variety of applications including NC, manufacturing planning, tooling and fixturing, inspection and Robotics. Each of these functions can access the product model and add their own results to the database. The foregoing concept describes Computervision's Designer System philosophy. (See Figure 3.1)

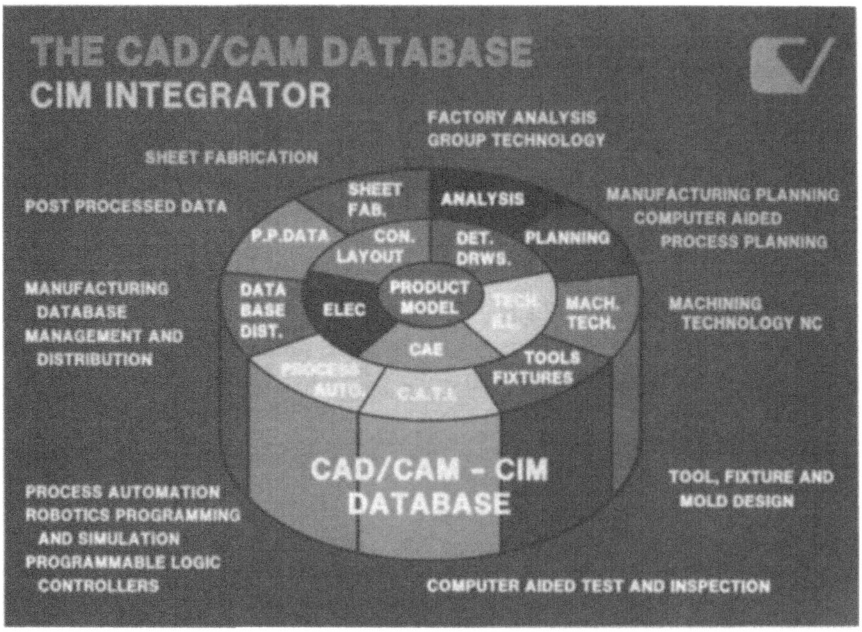

Figure 3.1. CAD/CAM DATABASE - CIM INTEGRATOR

Robotics, including workcell design, programming and simulation, is a natural CAD/CAM, CIM expansion area. Here the availability of a common engineering and manufacturing database is a major advantage. For example, a robot workcell may have NC machines or automated inspection equipment. CAD/CAM can program and simulate these automated machines along with the robot. Accomplishing these tasks, along with the multitude of workcell design activities, on a single integrated system results in many benefits. Section 5 expands on this.

4. ROBOTIC SYSTEMS

The heart of a robotic automated workcell is the robot. The major components of a robot are discussed here as well as the technological trends that have resulted in the marriage of CAD/CAM and Robotics.

4.1 The Controller

The controller orchestrates the actions of the robot. Using different methods, the controller stores the sequence of moves or operations. It can also provide for the interfacing with the robot's external environment.

Robot controllers have evolved in sophistication from mechanical, to pneumatic, to electrical, and finally to computerized forms. In each evolutionary step, the controller is able to control the robot for smarter and more complex tasks. Computerized controls also provide increased programmability. These controllers allow robots to perform a wide variety of manufacturing tasks in batch manufacturing. Examples include assembly, inspection and complex welding. Sophisticated controllers are a prerequisite for CAD/CAM graphics based programming.

4.2 The Manipulator

Robotic programming and simulation must consider the variety of robotic configurations. These include cartesian, polar, cylindrical and rotary. Each configuration has its own merits. For example, polar and cylindrical types can service a larger work space with a relatively small foundation. Cartesian robots need a large base and are easy to control and are well suited for applications such as spraying and welding. Rotary jointed robots provide good vertical reach, without requiring excessive space, and are quite flexible. Robots based on this type of configuration can reach over, into and overhead.

Each joint of a robot can have either a translation or rotational movement. The sum of these motion capabilities is referred to as degrees of freedom for the robot. For example, the Unimation Puma robot may be a 5 or 6 degree of freedom manipulator depending on the model purchased.

In general, the greater the degrees of freedom, the greater the robot's motion flexibility. With greater degrees of freedom the robot's cost also increases as well as programming complexity.

Finally, manipulators may be servo or non-servo controlled. Non-servo means limited sequence and implies simple controls and mechanical type programming. Motion is only possible by setting a limited number of mechanical stops. Servo robots can move along a continuous path or in a point to point fashion and require greater control and programming sophistication.

4.3 Robotic Programming Methods

Our definition of robotics included the term "reprogrammable." Programming a robot allows the implementor to put it to work for a variety of work tasks. The robotic programmer is concerned with defining the motion of the end effector or gripper. The gripper grasps parts and tools required to accomplish the automated task.

Non-servo limited sequence (pick and place) robots have the simplest type of programming, adjustable limit stops. Combined with Programmable Logic Controllers, these robots can sample inputs and outputs in the environment and respond with limited flexibility. Since this method of programming requires the use of the robot and its work environment, we refer to it as online programming. The major disadvantage of online programming is that the robot is tied up and out of useful production during programming.

Programming servo controlled robots is more sophisticated. The controller is usually a computer or microprocessor. Here continuous path and complex point to point motions are possible. Two methods are in use today: the teach method and offline programming using a robotic language.

4.3.1 The Teach Method - The teach method consists of walking the robot through its paces. An operator does this with a teach pendant, a box with switches that control joint or arm movements. Control of joint motion is very common. The more sophisticated robot controllers allow movement in robot world coordinates, in addition to joint coordinates.

An operator manually positions the robot gripper and records its position. Repeating this operation for a sequence of positions, the operator creates a program that moves the gripper from one position to the next in the given sequence. The limitations of the teach method are:

(A) The robot must be taken out of production in order to program. This means that the investment is not earning a return during off line time.

(B) Programming is tedious if many points are required. Such a condition frequently exists in aerospace and automotive applications where surfaces must be followed.

(C) There is no way to predict motion between recorded points. The intermediate motion requires the solution of the robot mathematics for motion prediction. When the robot is cycled the first time, the operator can be in for a surprise!

4.3.2 Offline Programming With a Language - The development of robotic languages such as RAIL (Automatix, Inc.) and VAL (Unimation, Inc.) are a big step forward for programming servo controlled robots. The programmer creates a computer program in an anologous manner that the NC programmer creates an APT program to control an NC machine. This robot program can be input to the controller and complex robotic operations performed.

Offline programming does relieve some of the need to tie up the robot for prove out, but certainly does not eliminate the prove out requirement. The limitations of offline programming with a language are:

(A) Offline languages provide very limited means to input the geometry of the workplace environment. This includes the part(s), or tool(s) for the robot to grasp, part presentation devices, fixtures, material handling equipment and machines.

(B) The programmer has no means to visualize the robotic motion and the interaction of robot motion with the rest of the workcell environment.

(C) The offline programmer is never really sure what his program will do on the factory floor. No programmer is perfect! The debugging phase ties up expensive manufacturing time.

Despite the above limitations, offline robotic programming is a major step forward. It is a prerequisite milestone for CAD/CAM graphics programming of robots, as we shall now see.

5. CAD/CAM ROBOTIC APPLICATIONS

The marriage of CAD/CAM and robotics results in increased productivity for robotic implementation activities. CAD/CAM robotic applications include:

(1) Robotic workcell design
(2) Robotic workcell programming
(3) Robotic workcell simulation

5.1 Robotic Workcell Design

The robotic workcell contains all the physical equipment needed to create a functioning robot application. Besides the robot, a workcell can have special fixturing, other automated machines (NC machines, Coordinate Measuring Machines, visual inspection equipment, etc.), material handling devices, part presentation equipment, robot grippers, etc.

The equipment in the workcell must be arranged in such a manner that the robot work envelope includes all required device areas. CAD/CAM is perfect for designing the equipment layout. Different configurations can be tried to find the optimum workcell configuration.

Libraries of workcell components can be stored on the CAD/CAM system and recalled when needed. For example, a robot library could contain commercial robots along with their work envelopes. Libraries for standard fixturing, machines and other workcell equipment are also recalled from the CAD/CAM system.

All the workcell elements are designed with CAD/CAM. With a well integrated CIM concept, the groups supporting the workcell implementation use CAD/CAM to design and manufacture workcell elements.

CAD/CAM workcell design results in many benefits. First, the design activity is more productive. This means that design time and costs are reduced. Second, CAD/CAM for workcell layout allows more alternatives to be considered resulting in an optimal layout. Third, the lead time to design and layout the cell is reduced. This contributes to an overall reduction of robotic workcell lead time. Fourth, the quality of the designed components and the overall cell is cincreased.

5.2 Robotic Programming and Simulation

The infusion of computer technology in Robotics, coupled with the move into batch manufacturing, led to development of offline robotic programming with a language. The next step in the evolution of robotic programming is graphic robotic programming and simulation. First some definitions are in order.

Graphic robotic programming is defined as interactively specifying the trajectory of the robot's end effector. Model geometry in the workcell environment is used to help specify this path. The programmer uses both interactive and automatic techniques to specify the end effector's X,Y,Z position and the orientation of a control point axis and its twist about that axis at any point along the path.

Graphic robotic simulation is defined as simulating the programmed robot path. The simulator contains the joint mathematics that fully describe the target robot. Simulation checks whether or not the robot can position its end effector to the positions and orientations specified during programming. It is always possible that the target robot's end effector cannot assume the desired position and orientation required. Revising the path or workcell may be required in this case. Also, the robot's degree of freedom may not be sufficient to accomplish a given task.

Simulation, then, creates the actual robot trajectory using end effector positions available from graphic robotic programming. The actual robot motion between these positions may deviate from what the user desires. This is caused by the robot's interpolated joint movement between points. This deviation can be lessened by programming the position points closer together. Some robots, such as the Unimation Puma and Automatix AID 600, can move in a linear fashion between points. This will also reduce the discrepancy between actual and desired trajectories.

Graphic robotic programming and simulation result in a number of key benefits. These benefits relieve the deficiencies in offline programming with a language noted in section 4.3.2.

5.3 Robotic Programming and Simulation Benefits

Graphic programming is accomplished using the complete robot workcell geometric model. This model includes the part(s) and tool(s) to be manipulated along with models of all other equipment in the cell. Programming in a known and accurate model environment reduces programming time and errors.

As the robot gripper is programmed along its trajectory, simulation provides the user with accurate visual feedback and verification of the operation. Just compare the efficiency and reliability of blind offline programming using a language with sighted programming using graphics and the vast advantage of a CAD/CAM approach to robotics becomes apparent.

Graphic robotic programming is safer than other methods. Manipulator programs are debugged using simulation without endangering arms, workpieces, fixtures, or factory personnel. When the robot program is run in the actual workcell, prove out time is significantly reduced, hence the robot has more productive up time.

Graphic programming can output the target language or input for a particular robot. This means that CAD/CAM offers a common method to program different makes of robots. The programmer does not have to be an expert in multiple robotic languages. Programming on the CAD/CAM system reduces the interactive demands on the individual robot controllers which in turn reduces the costs of the controllers.

6. COMPUTERVISION'S ROBOGRAPHIX™

A major step forward in the evolution of CAD/CAM and Robotics was taken with the announcement of Computervision's Robographix product. Robographix is the first graphic robotic programming and simulation product that is totally integrated in a turnkey CAD/CAM system.

The following activities can occur on Computervision's CAD/CAM Designer System using a common engineering and manufacturing database.

(A) Engineering designs the product, components, or parts that are manufactured with the robot.

(B) Process Engineering can layout the robotic workcell and evaluate alternative robot selections, cell configurations, etc.

(C) Tool Engineering can design all fixturing and special equipment needed for the workcell.

(D) The robot can be programmed and simulated with Robographix. The workcell model is used here.

(E) Other automated equipment such as NC machines and Coordinate Measuring Machines can be programmed which are required in the workcell.

(F) The machine programs for the robot, NC machine, etc., can be DNC communicated from the CV Designer System to the workcell.

6.1 Robographix Overview

Robographix is capable of programming and simulating robots that have intelligent program input. This would include such robots as the Unimation Puma series with the VAL language, Automatix AID series with the RAIL language and other appropriate robots.

Since Robographix is integrated with Computervision's Designer System, the results of all upstream robotic workcell design activities are available to the robotic programmer. The programmer uses the workcell model, as well as the part or product model during the programming session. If the robot must interact with other NC or inspection machines, these too can be programmed.

Robographix is a generalized robotic programming and simulation tool for a variety of robot tasks including assembly, welding, material handling, machine tending and spraying.

Robographix is available on Computervision's CGP 200X CADDS 4X Designer System. It can be used in conjunction with CAMACS, a DNC software communication product, to distribute the robotic program to the workcell.

6.2 Robographix Features

Robographix consists of two parts:

(1) Robot path generator
(2) Robot Description and Language Processor (RDLP)

The robot path generator is the same for any robot. It generates the generic portion of the robot path. The RDLP is specific for the target robot. It contains all the joint mathematics that describe the robot. It also contains the rules for transforming the robot path into the language/input that the robot can understand. In this manner it is analogous to an Nc postprocessor.

Some of the main features of Robographix are:

(1) It provides interactive robot path generation using a menu of options.
(2) It provides up to 6-axis control.
(3) It references libraries of robots contained in the system.
(4) It provides the ability to enter any valid non-motion robot command.
(5) It detects and inhibits joint limitations.
(6) Joint positions and link geometry can be statically displayed.
(7) The full robot path can be dynamically simulated.
(8) The tool control point can be programmed with offset.
(9) Different robots can be tested on a single path.
(10) Point to point and continuous robot paths can be programmed.
(11) The robot path is processed to create the actual robot input.

7. CAD/CAM AND ROBOTICS EXAMPLES

The following examples demonstrate the synergy of Robotics and CAD/CAM. Computervision has promoted the evolution of Robotics in a CAD/CAM environment and demonstrated this commitment at the Autofact III and IV shows. (November 1981 and 1982). The robotic workcells were designed on CV's Designer System and Robographix was used for robotic programming and simulation.

7.1 Automated Manufacturing Cell at Autofact III, 1981

The part manufactured at Autofact III was a medallion (Figure 7.1.2) with a number of complex machined features. The workcell consisted of a Unimation Puma 600 robot, a Pratt and Whitney three axis NC milling machine and a Bendix Cordax coordinate measuring machine. Figure 7.1.1 shows the actual workcell.

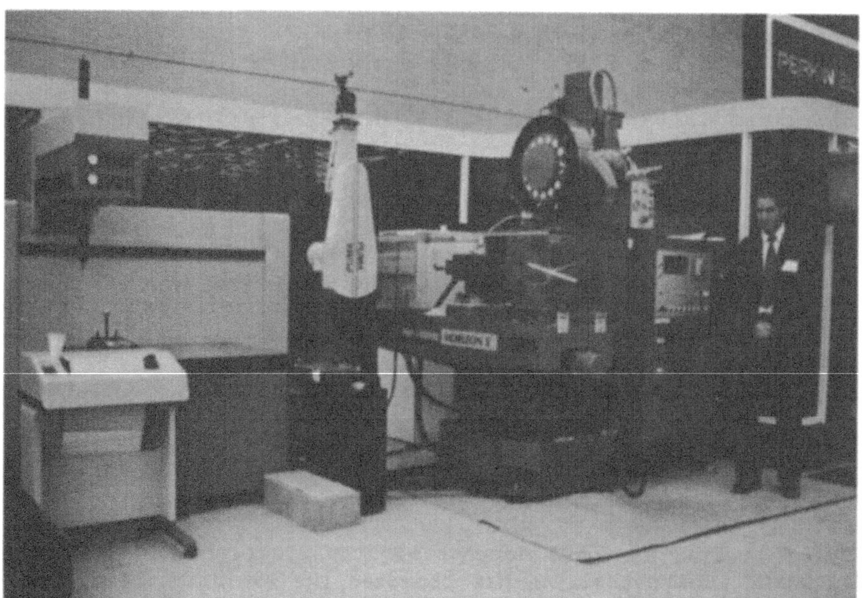

FIGURE 7.1.1. AUTOMATED WORKCELL - AUTOFACT III

FIGURE 7.1.2. THE MEDALLION

The work sequence began with the robot picking up an aluminum blank from a magazine at its base. The robot then placed it in an automatic fixture on the milling machine. The blank was then NC machined. The robot unloaded the milling machine and placed the finished part on the coordinate measuring machine for quality inspection. This completed the automated cycle.

Figure 7.1.3 shows the layout of the workcell on the Designer System. In addition to the medallion design and workcell layout, the Designer System was used for fixture design and process planning. The robot, NC machine and coordinate measuring machine were programmed graphically. Figure 7.1.4 and 7.1.5 show part of the Robographix simulations of loading and unloading the NC machine and placing the part onto the inspection machine. The Robographix simulation proved out the robot path and verified that there were no collisions.

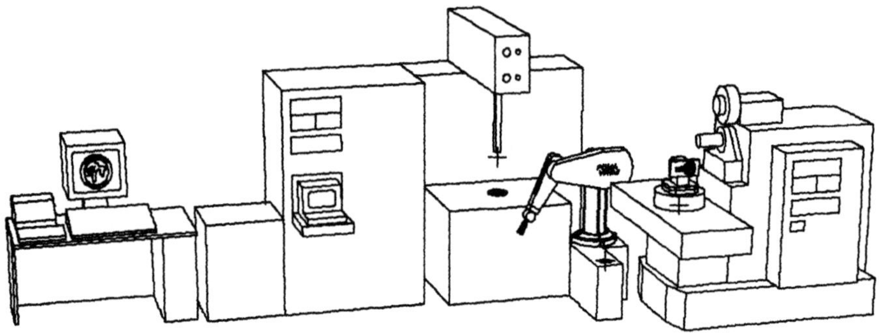

FIGURE 7.1.3. WORKCELL CAD/CAM LAYOUT

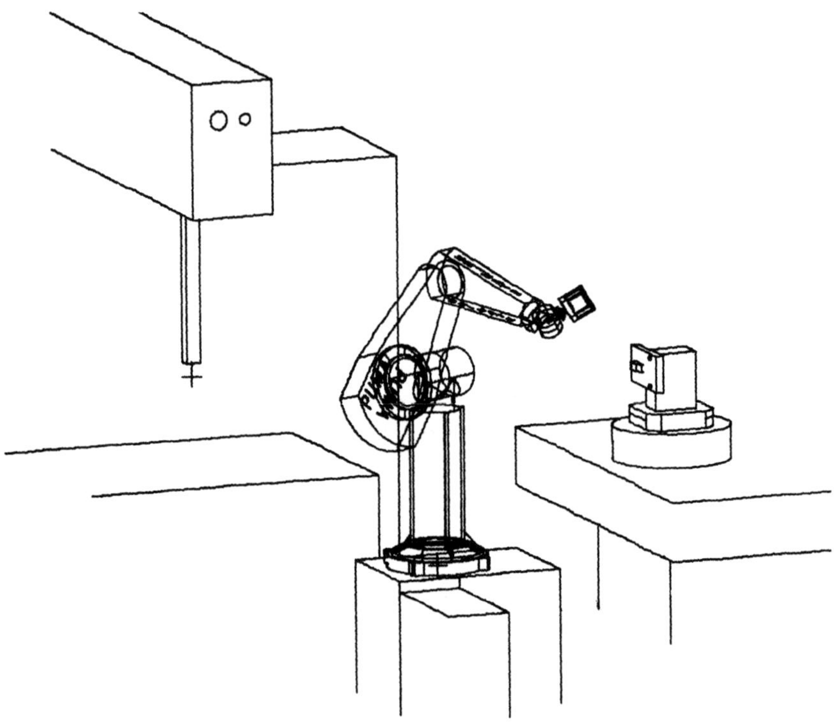

FIGURE 7.1.4. ROBOGRAPHIX SIMULATION OF PUMA ROBOT LOADING THE FIXTURE

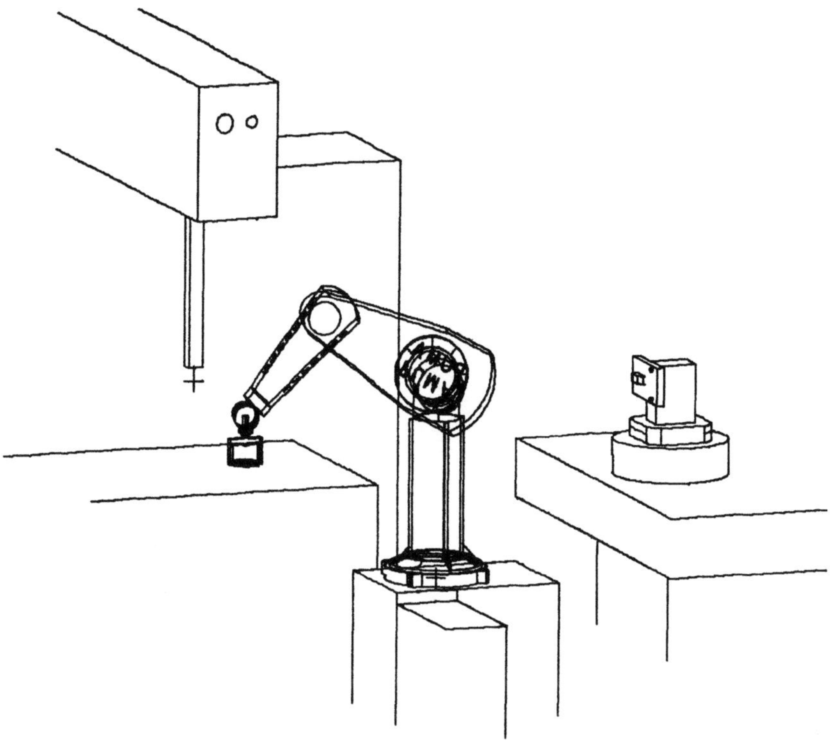

FIGURE 7.1.5. ROBOGRAPHIX SIMULATION OF PUMA ROBOT PLACING PART ON COORDINATE MEASURING MACHINE

7.2 Automated Sheet Metal Manufacturing Cell at Autofact IV,1982

Figure 7.2.1 shows a robotic workcell that was created to manufacture a sheet metal part at the Autofact IV Show. The workcell consisted of an Automatix AID 800 robot, a Wiedematic NC turret punch press, and an Automatix visual inspection station.

FIGURE 7.2.1. AUTOMATED SHEET METAL WORKCELL AT AUTOFACT IV

In this demonstration the robot began the work cycle by using a suction gripper to pick up a blank sheet of steel from a magazine loading table. The robot placed the sheet on the NC punch press. The sheet was punched on the machine. At the NC machine's home position, the robot marked an identification number by rotating its gripper to position a marking tool. After completing the marking operation, the robot unloaded the punch press and deposited the completed sheet on a visual inspection table. After inspection, the robot placed the good sheet on a finished part conveyor.

Figure 7.2.2 shows the robot cell layout that was designed on the Designer System. Robographix was used to program all robotic movements. The simulations (Figure 7.2.3) proved out the overall robot motions and checked for possible collisions. The NC turret punch press was also programmed using Computervision's Autopunch. The machine programs were DNC communicated to the workcell using Computervision's CAMACS.

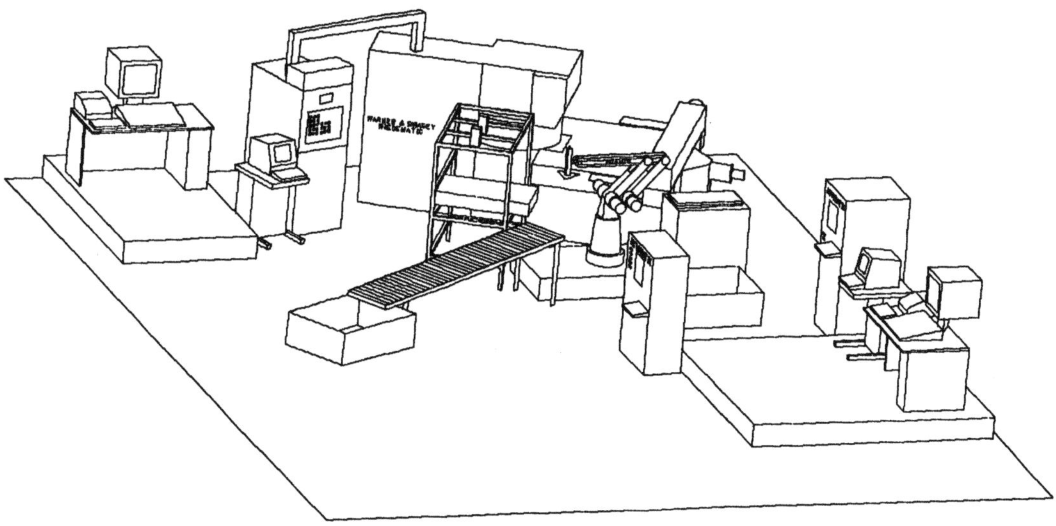

FIGURE 7.2.2. WORKCELL CAD/CAM LAYOUT

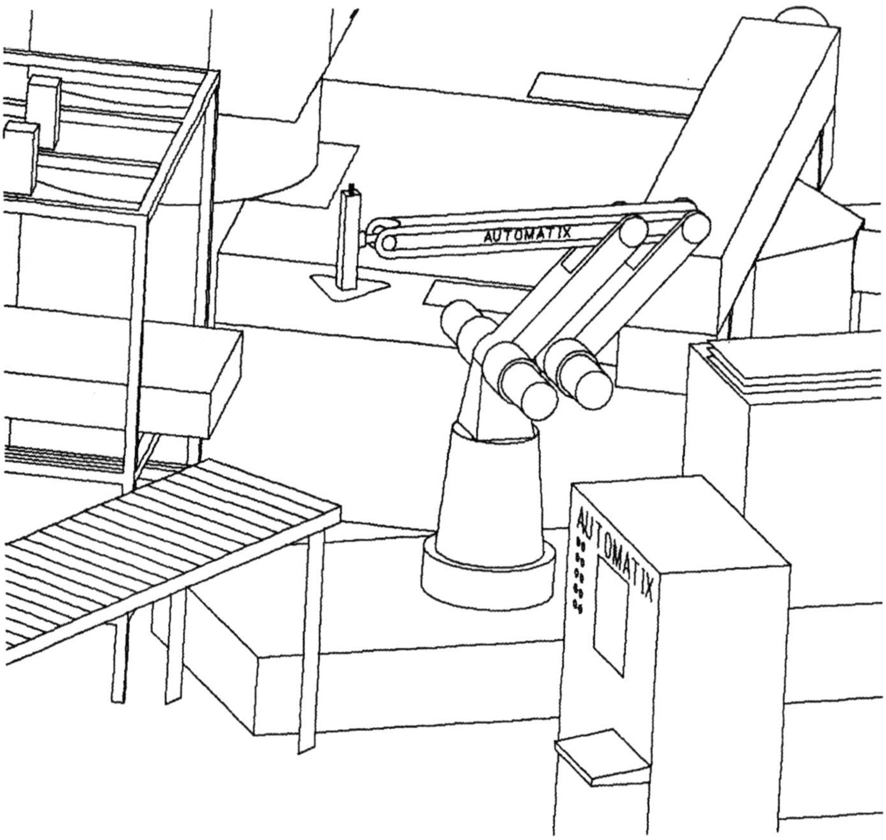

FIGURE 7.2.3. ROBOGRAPHIX SIMULATION OF AUTOMATIX ROBOT

8. SUMMARY

The third Industrial Revolution starting with the introduction of CAD/CAM in manufacturing has only begun. The 1980's will see CAD/CAM become an established necessity for successful implementation of robotic applications in batch manufacturing. Products, such as Robographix, will be expanded to include greater levels of simulation for the entire workcell environment. This will include the programming and simulation of inputs and outputs for nonrobotic devices and sensors. CAD/CAM links will be made for vision applications. Solids modeling of the workcell environment will allow for automatic collision detection and avoidance methods. Solids will also advance the state of the art in graphic robotic programming.

So, the manufacturing technology evolution continues. The marriage of CAD/CAM and Robotics has resulted in a great synergy. This synergy will help make possible the move of robotics into batch manufacturing.

9. Biography

Mr. Simon is Director, Mechanical Applications Product Marketing and is responsible for all applications and product planning that impact the Mechanical CAD, CAE and CAM market areas. He is the author of a number of technical articles appearing in technical journals and the Machine Design Magazine Design News on CAD/CAM subjects. He is the holder of a US patent. Mr Simon received a BS and MS in Mechanical Engineering from Rensselaer Polytechnic Institute and has a Masters of Business Administration from the Boston University.

References:

Simon, Richard L., "CAD/CAM Foundation of Manufacturing Automation," Conference on CAD/CAM Technology in Mechanical Engineering, March, 1980, pp. 221-235.

Cook, Craig, "Robot Programming in a CAD/CAM System," 12th Annual International Robotics Symposium.

Soroka, Barry, "Debugging Robot Programs with a Simulator," CAD/CAM 8 Conference, Anaheim, CA, November, 1980.

Heginbotham, W.B., "Computer Graphics Simulation of Industrial Robot Interaction," Proceedings of the 7th I.S.I.R. 3rd C.I.R.T., Tokyo, November 1977.

24

A LOOK AT TECHNICAL DOCUMENTATION AUTOMATION AND ITS PLACE IN DISTRIBUTED DATA BASE INFORMATION SYSTEMS

by Richard C. Sisman
Management Consultant

Abstract

Developments in technical documentation automation have continually improved the process of packaging text, tabular, and graphic information. Systems that utilize batch, interactive or combinations of each method of processing information into a deliverable medium are available, some providing integration of graphics with the text and tabular matter. There is a continually growing interest in the writer's workbench concept, as there is in programmer's and engineer's workbench transpirations.

Technical documentation automation has developed more rapidly and is being used more extensively that distributed data base information systems, to which it must ultimately be integral. Some users of automated technical documentation systems presently share data among other closely related technical document preparers (i.e. training, technical manuals, maintenance engineering) creating a very limited distributed data base. However, the potential facility and thus productivity will not be reached until those support entities share the data bases of operations (where applicable), engineering, and manufacturing. Only then can all relate, each to the others, and interact when information is created, changed, and used.

PREFACE

Keynote speakers at the Defense Computers-Graphics conference (DCG-84 in Washington, D.C. last year* unknowingly set the theme for this and other papers - perhaps this entire conference. Commodore Grace M. Hopper said, "The Model Ts of the computer are here .... We have been through the preliminaries," Commodore Hopper said we should now focus on the information. She said we have spent 40 years looking at how we process it and we now have to look at the data itself, the output, and at the flow of information. She went on to also stress we are not using the best methods of presenting information - particularly graphic methods.

Deputy Under Secretary of Defense for Research and Advanced Technology, Dr. Edith W. Martin pointed out we understand the magnitude of the computation challenge and have begun to develop and implement our response to that challenge. "Graphics," Dr. Martin said is "... the next real challenge ... (and) the bottleneck of 1990s if nothing is done right now." Dr. Martin also points out that graphics cannot be treated as a separate entity, but must be considered along with the total future computer/information requirements.

The attendees and the speakers at this conference understand the problems, the requirements, and the challenges. All share the same concerns, although they perhaps are directed toward more specific areas of interest or responsibility. This paper, like many of the others treats a specific area, technical documentation automation, yet still will address the larger issue - information, its generation, processing, change, flow, and use.

INTRODUCTION

The creation, change, and use of technical information has long had the attention of those who know the value of applied automation technology. In the late 60s there was a project at Hughes Ground Systems Group to develop a computer program to take signal, wiring, component, and logic information and create a logic mechanization diagram complete with signal lists and input/output tables for output on a line printer. There were a couple special characters on the line printer chains to accommodate flow lines and connections - and it worked satisfactorily. The program was later enhanced to relate a change in one element to other elements affected by the change and then make all the changes. Further enhancements included the addition of an output to drive a third generation phototypesetter for better quality drawings. The end result was very good technical documentation produced by an automated system.

*As reported in "Computer Graphics Today," October 1984.

Unfortunately the potential of that program was not exploited beyond the mechanization of logic in a circuit board environment. Many engineering operations could have them and still need automation applied in this manner. This concept, formulated over fifteen years ago, is still valid. The concept evolved from a need to more easily create, change, and use information, not from the need to automate functions. However, the reality is that we automate functions to accommodate creation, change, and use of information.

Functions such as; word processing, computer-aided drawing/ design, local and wide area networking, storage and retrieval, all the related technological areas have developed, although many are still, as Commodore Hopper put it, "Model Ts." It is this world of **functions** that this paper will concentrate on - automated technical documentation.

TECHNICAL DOCUMENTS

A list of technical documents must certainly include: engineering drawings; specifications; standards; various lists such as wire, parts, signal, etc.; logic mechanization; programs and programming information; instructors guides; student training literature, technical manuals, maintenance analyses; proposals; reports; and on-and-on.

Figure 1 shows the technical documentation in a large electronics company. Interrelationships are also shown for when we discuss documentation in a distributed data base information system. The form and format of these documents are largely determined by the users requirements, although they may also be determined by the capability of the output devices being used. Documents used internally tend to be of lower quality (form and format) than those delivered to other organizations or to the customer. Assuming a moderate level of automation, line printers for text and tabular matter and electrostatic plotters for graphics usually provide satisfactory output for internal use. Pen plotters, letter quality printers, and even phototypesetters are most frequently used to output documents for external distribution.

SIMPLIFIED AUTOMATED TECHNICAL DOCUMENT FUNCTIONS (See Figure 2)

For those whose job it is to formalize the information being prepared and processed in operating and support organizations, the functional requirements are simple and straight forward. As shown in Figure 2, the text and tabular information needs a word processing function and graphic information needs a graphics processing or automated drawing function. Text, tabular, and graphics are input either from external sources (other organizations, other systems) or directly at local input

FIGURE 1. INTER-RELATIONSHIP OF ENGINEERING, FABRICATION, SUPPORT DATA

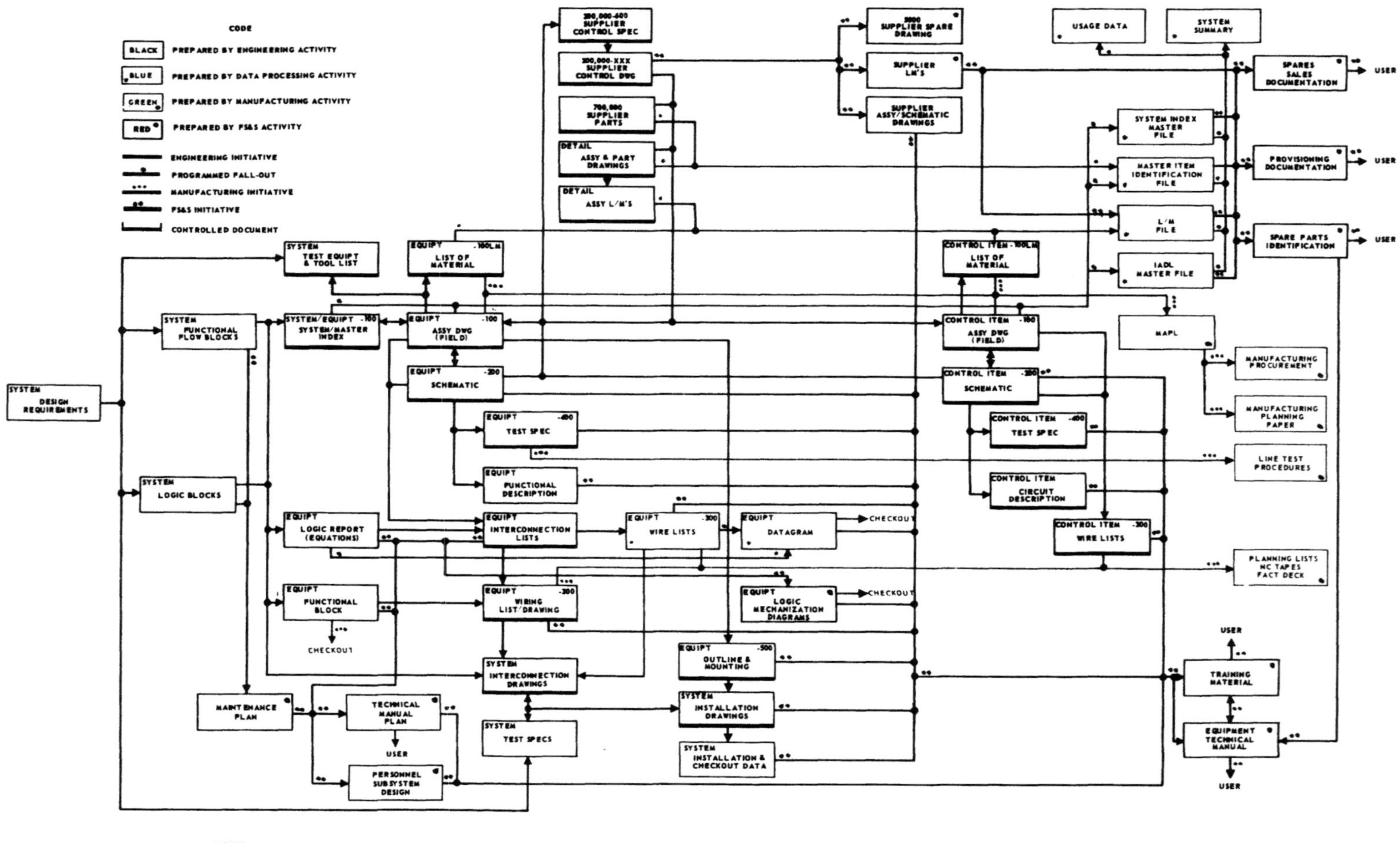

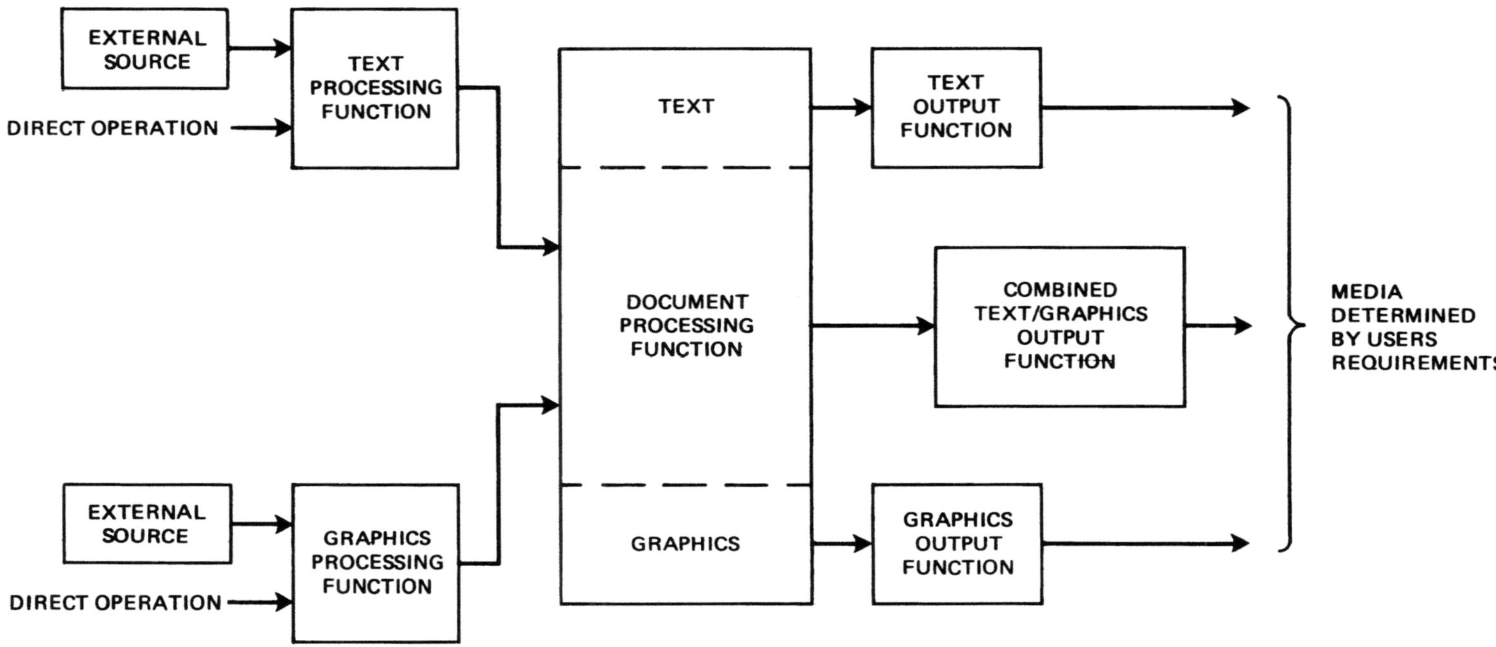

Figure 2. SIMPLIFIED AUTOMATED TECHNICAL DOCUMENT FUNCTIONAL DIAGRAM

stations. Inputting from external sources will often pose problems of communication and translation that are not present in direct operation.

The key function is that of document processing which may be an integral part of a larger system, a node, or a standalone capability. It is software intensive and how automated the process is depends on the subfunctions available to provide documents to meet user requirements. Beside formatting the document, the document processing function will, for text; paginate, position footnotes, structure equations, select typefaces and sizes, place heads and feet, hyphenate and justify, rule tables, insert revision bars, prepare an index or table of contents, allow graphic space or insert a graphic, and more. For graphics, the document processing function is limited to some formatting and sizing subfunctions, and to type selection or substitution. However, output format requirements are not as demanding for graphics as they are for text.

Output functions most commonly accommodate text and graphics separately thus requiring some manual intervention. The more sophisticated systems combine text and graphics either in the document processing function or in the output function. Media provided by the output function is typically paper or microfilm. There are efforts directed toward providing output in digital electronic form for either on-demand printing or a CRT type user device. Since these latter efforts have abounded for the past ten years, it is possible they may finally make an appearance in the near future.

When structuring this type of automated technical documentation system it is necessary, when analyzing the requirements, to decide how much "man" and how much "machine" you want on the system; whether you want your system to be an interactive (for document processing), a batch system (letting the computer make the decisions), or some quanity of each. Document format or structure requirements should be the principal deciding factor. The majority, over 90 percent, of technical documents have set structures that can be programmed into a computer. The system can make consistently correct decisions, and direct a device to output the document without a "man" in the process. For the less than 10 percent of documents that require "mans" esthetic touch or communicating skills, and for those who still do not trust a "machine", the interactive system must be considered. In larger operations with a variety of document structures and types, some of both probably can be justified. I must point out, however, that the principal benefit of automation, and how most of us justify the capital expenditure, is the reduction in manual labor. A detailed analysis of capital versus operating expense in both the short and long term must be made to make the decission. For example , on interactive systems - may be cheaper to buy but more expensive to use than a batch system.

A CASE STUDY - DOCUMENTATION AUTOMATION

One of the companies I have provided consulting service to, Informatics General Corporation, Publishing Services Division, is interesting case study of documentation automation evolution. As basically a software company, when the need to sophisticate the documentation required by Informatics' own operation, and as a demand developed from customers, they decided to write their own. The software was developed by individuals with expertise in publishing disciplines. It was a powerful batch program that could compose the pages of complex legal documents as well as a wide variety of other documents that are normally best presented when typeset. To input the Informatics system, only a few generic codes are needed instead of the extensive (output) device dependent codes previously required by systems. The generic codes identified the generic parts of the document (ie, paragraph, table, heading) and from these the program can create the form and format of the document and then insert the device codes needed to drive the particular output device being used. As in all such stories, the demand to provide service increased and interest in the system grew to a point that it was decided to refine and then market the software. Informatics Composition System IV (CSIV) software designed to run on IBM mainframe and output to a third generation phototypesetter was the first package to be generally marketed, and it is still available. CSIV has all the composition and pagination functions, including text/graphic merge from either a scanner or from a CAD system, that can be put on a system. It is a complete batch system.

The next effort was to rewrite CSIV for a smaller general purpose computer. The Wang VS series was selected because it provided a good architecture, was relatively inexpensive, and came with one of the best word processors on the market. There were always drawbacks in using the IBM mainframe. Among them was the IBM computer usually belonged to someone else, frequently a problem for other users, and to get maximum effectiveness from the system a separate good front-end word processor was needed, often a Wang. So, Composition System V (CSV) was developed. Both CSIV and CSV do exactly the same things and one or the other is in place at dozens of companies like Federal Express, Pratt and Whitney, Prudential, Mathew Bender, FMC, and AT&T.

The next phase of product evolution by Informatics was to package a system called Info Page. It is composition software, a small Wang VS computer, and a phototypesetter. This provides turn-key capability for those who have a relatively low volume of documentation at an affordable price. Some features of the software, notably text graphic merge, are not included in this product.

The most recent offering by Informatics is the addition of interactive text and graphic page composition capability using the Camex "Breeze". This is available for both CSIV and CSV. The addition of "Breeze" provides interactive capability for users in the predominantly batch environment of technical documentation.

It should also be noted that, along the way, Informatics answered the need to output to other than phototypesetters. Drivers are available for other output devices including laser printers. This is important since laser printers are a likely candidate to meet on-demand printing requirements in addition to being a replacement for phtotypesetters as their quality (resolution) improves. All-in-all systems, such as those offered by Informatics, become key elements in the technical documentation element of distributed data base information systems (DDBIS). Another related development is that taking place for the writers workbench concept.

WRITER'S WORKBENCH

The concept of a writer's workbench is not new. There have been some notable efforts. One, the Bell Laboratories collection of programs, provides an analytical tool for the writer to use to better perform his job. Its concentration is on analysis, reporting the analysis, and, in some cases, suggesting changes. In addition to the analytical software, a writer's workbench must also provide integrated text, tabular, and graphic capability, and be a distributed data base system interfaced to other data bases in engineering, manufacturing and support operations.

The writer's workbench, as a concept, has probably been around since shortly after E. Remington and Sons produced the first practical typewriter in 1874. There is little doubt that some enterprising writer bought a typewriter, took it home, placed it on a desk and started using it to create his copy. The term "writer's workbench" probably wasn't in use in 1874, it is a more recent term, but none-the-less the typewriter was a basic writer's workbench that could be used to enter copy and to type revised copy. It wasn"t until the advent of media assisted typewriters (usually tape) that changes to copy became as easy as original inputting. And, it wasn't until the computer was attached that we could reach the level of sophistication today's writer's workbench can provide.

Where technology, particularly advances in software, has taken us today, is exemplified by the developments at Bell Laboratories in their version of the writer's workbench. Bell Labs has developed a set of computer programs to help writers produce better documents by analyzing the document, reporting on significant elements, and providing, in some cases, suggested changes.

The first group of Bell Labs' writer's workbench programs are designed to assign word classes to English text - to find the parts of speech. The group of four programs is called PARTS. Among the things the PARTS system does, in addition to identifying the words and sentences and then assigning classes (i.e. noun, vert, etc.), is to produce an outline from headings and even create a rough abstract from topic sentences, when they are used in the text. When tested, PARTS was found to be 95 percent accurate on both graded and technical text.

With the output of PARTS it is possible to analyze writing style, compare the writing styles of documents, and direct rewriting as required using other Writer's Workbench programs. Among these are STYLE, PROSE, REWRITE and TOPIC.

STYLE is a program that uses PARTS output to measure some of the surface features of the text. The first section of STYLE outputs four readability indexes. These are Kinkaid, Coleman-Liau, Flesch, and the Automated Readability Index. The second section provides sentence information such as: number of sentences and words; average length of both; number of questions and imperatives; number, percentage, and average length of non-functional words; and details on sentence lengths. The next section shows the number and percentage of each type of sentence (i.e. simple, compound, complex, compound-complex). The fourth section outputs the number and percentage of each word type and the final section provides number and percentage of the various types used as sentence beginnings. STYLE provides a relatively complete analysis of the style and complexity of the text from which problems can be identified and corrective measures taken.

The next program used by Bell Labs, PROSE, takes some of the statistics collected on a document by STYLE and compares them to standards derived from other similar type documents. The output describes how the document compares to a "good" document in readability; sentence structure and variation' and use of passive verbs, nominalizations, and expletives. PROSE then gives the writer suggestions for improving the document.

REWRITE is a program that finds sentences with passive verbs, many extra words, and long strings of prepositional phrases and then lists them as candidates to be rewritten. It uses both the output of PARTS and a copy editing program, DICTION, discussed later in this paper.

TOPIC uses PARTS to locate frequent noun phrases in a document. Since the noun phrases used most frequency in a document are usually the topics of the document they may be used as key words or indexes.

Bell Labs uses several programs as copy editing programs. The first of these, DICTION and SUGGEST prints all sentences in a document containing phrases that are frequently misused or suggest wordiness and, then when prompted, will print an alternative. It is basically a string matching program that can be used with different data bases to find other kinds of phrases or words. In this case the data base has about 450 phrases as a default pattern file. The Bell Labs writer's workbench also uses data bases of sexist terms and of abstract words to be run by DICTION.

Other copy editing programs are SPELL, DOUBLE, PUNCT AND SPITLINF. These programs do about what their names imply. Check spelling, look for consecutive occurances of the same word, look for simple punctuation errors, and find and provide grammatical information about split infinitives.

There is more, but this gives an idea of how far this aspect of a writer's workbench concept has been taken. Bell Labs is not alone. There are developments by others. Among these are IBM's EPISTLE; University of Michigan's JOURNALISM, obviously for use by journalism students; the U.S. Navy CRES for use in increasing the quality of technical manuals and training materials, and the Westinghouse Writing Aids System. It should be understood that these, along with Bell Labs writer's workbench, are all basically analytical tools to report on the "Quality" of the material and thus help improve the writing. The concept of writer's workbench that is considered herein is much broader in scope.

A writer's workbench, particularly one to be used in the technical publications environment, must include some or all of the above analytical software; it must also be interfaced to all engineering, some manufacturing, and all support operations data bases; and it must support text, tabular, and graphic information.

There have been some earlier efforts in other areas that parallel what many are doing in developing the technical writer's workbench. The most advanced of these is the Programmer's Workbench that has been in use for several years. Simpler, because it does not require much, if any, graphics capability, the programmer's workbench does provide a programmer with the tool to create, analyze, test, and document programs. The other major effort at a workbench concept has been in engineering. In addition to what has been done in CAD/CAM, many companies are beginning to put together data base systems using an engineer's workbench with text, tabular, and graphic capability. The objective is to give the engineer, as with the programmer, a means to design, analyze the design, test, and document the design. In both the programmer's and engineer's workbench development, the interface and interaction with other data base systems has been a limitation, but gradually is being overcome. Another limitation in the engineering workbench development, as it is in the writer's workbench development, is integrating text and graphics. Presently each is usually done separately on separate systems and changes on one must be made on the other. What is desired, and in some areas being accomplished, is full integration. If a change is made to the drawing, such as adding a component and modifying the wiring, that change can be automatically reflected in other documentation (i.e. parts list, wiring lists and diagrams, specifications, etc). In addition, it might also be necessary to know the impact on other parts of the design.

Therefore, the first requirement of a Writer's Workbench is that it be the access to a publications data base which is interfaced to other data bases. The need is for this total structure to be a group of integrated distributed data bases with common management. The publications data base must interface to other support data bases (i.e. training, provisioning, maintenance engineering, etc.), and to data bases within the engineering and manufacturing functions.

Other basic requirements of this concept of a writer's workbench include:

1. It must be a word processor where we can create and change textual and tabular information.

2. It must be a drawing system where we can create and change graphic information.

3. It must integrate text, tabular, and graphics so if one changes it is reflected in the others.

4. It must be an analytical device to perform task pertinent analyses (i.e. Bell Labs).

5. It must process the text, tabular, and graphic information to a selected output medium with minimal man and maximum machine intervention.

The above are key basic requirements to create a totally functional technical writer's workbench. There are many trying to put parts of this into place - hoping to ultimately have it all. The author has been involved in similar efforts at Hughes Aircraft Company Ground Systems Group and Federal Express Corporation. The Federal Express Corporate Automated Publications System (CAPS) is shown in figure 3, with the writer's workbench highlighted. At present, the CAPS writer's workbench interfaces only with the technical publications, policy and procedure data bases. Interrelationship to engineering and training have also been considered.

TECHNICAL DOCUMENTATION IN DDBIS

Automation of text processing and document publishing systems are at a more advanced state of development, generally, than automated design and drawing systems and certainly of distributed data base information systems. Although the technological problems are being solved, the acceptance of these

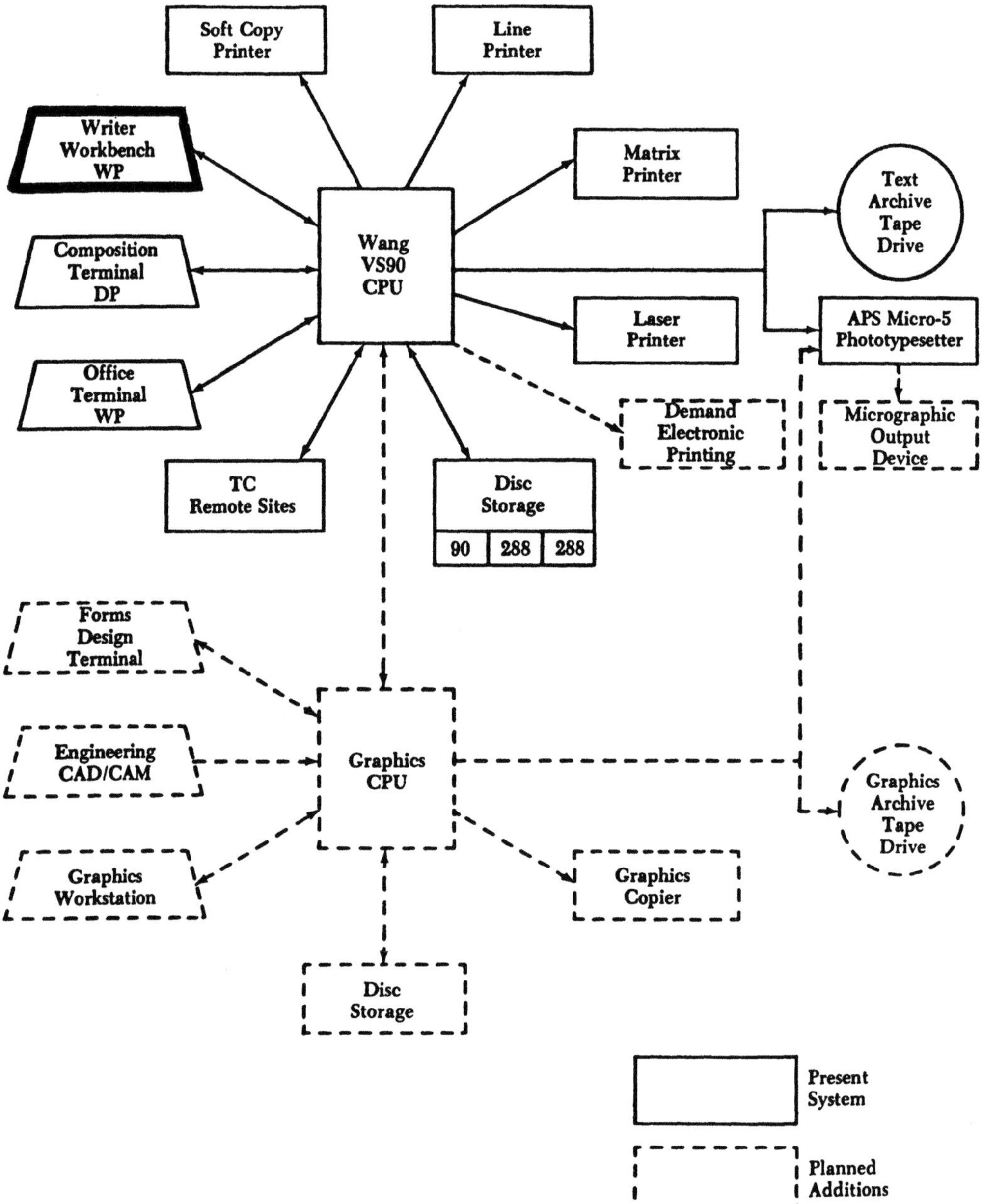

Figure 3. Federal Express Corporate Automated Publication System (CAPS)

concepts, their productivity impact on valuable resources, seems to have generated less enthusiasm than expected in upper management. Understandably, capital intensive programs move more slowly since economic factors have a substantial impact on capital expenditure. I have hoped that enlightment would come sooner.

Although Figure 1, presented earlier, is over fifteen years old, the documentation shown consists largely of the technical documents you find today in an electronics manufacturing company. It will also probably be the same fifteen years from now. What has changed is the manner in which we create and use the documents. Some documents are now the product of an automated function, such as schematic drawing; some groups of documents may be produced by a larger data base system, such as various types of parts lists; even some interrelational operations exist, such as CAD/CAM in circuit board or LSI environments. But total interrelationship and interaction is still a dream (or perhaps nightmare) to most of us. What we ultimately want is the distributed data base information systems. We want a system with the intelligence to interrelate changes and interact with all elements that change. We want a data base that allows cross use of data and reporting of information. Although we are here concerned with technical information and the class of documentation shown in Figure 1, interrelationship and interaction must include program management, resource management, and even general operations to some extent. We need to know when and how an engineering action impacts manpower, material, money, schedule, etc. We also need to know for example, the impact on other elements of a manpower decision or action. We need to have all of this for maximum effectiveness. When can we have it? Probably not, to any large extent at least, in the near future. Technological areas of communications, artifical intelligence, relational data base structures, storage, etc., may permit a modest beginning but a large scale effort needs some technologies to be further refined and a lot more dedication to the concept than has been exhibited to date.

But, there are other questions. Do we really want it at the probable cost? Is the benefit balance largely to management in their decision making? Thusly, can significant productivity be expected in development, design, manufacturing, and support operations? Questions logically follow as to the scope of the task. Is it too large an undertaking? Can the effort be segmented and performed piecemeal in segmented steps over a long period of time? Can performing only parts be an acceptable solution? Have we learned how to perform such a task that crosses organizational/functional perogatives?

Having looked at this concept while still at Hughes, having considered the related problems, having sized the task, and having put it in some time frame of accomplishment it was clearly going to be a challenge. It was then I decided to retire. Not true, but it was, and is, an undertaking that needs both philosophical and staff continuity over a long period of time. Since it cannot be started fresh by throwing everything that exists away; the details of what is there, what is added, and what is changed need careful complete study and documentation (the subject of this paper). Therefore that element of the system should be among the first developed and refined. So, whether as a separate function or an integral part of a grand DDBIS scheme, automation of the technical documentation function is, obviously, an important and necessary requirement.

In conclusion, it can be stated that the means to automate the creation or preparation of text, tabular, graphic, or combination technical documents now exists in a relatively refined state. Interrelating the information elements to fully facilitate use and revision of the information is less refined as it relates to a distributed data base information system. When the need is accepted as valid, the resources will be found to perform the task. So it has been!

BIOGRAPHY

RICHARD C. SISMAN is a management consultant in the fields of automated publication/publishing and information resource management. He is retired from Hughes Aircraft Company GSG after 25 years in various management positions, most recently as Manager of Management Systems. Prior to Hughes Mr. Sisman was an educational specialist and supervisor of technical training for the USAF Air Training Command. He has been active in several associations including the National Security Industrial Association and the Aerospace Industries Association. Among the corporations Mr. Sisman has been a consultant to are Informatics General and Federal Express.

25

SUCCESSFUL BUSINESS COMPUTER GRAPHICS

Susan W. Ulicny
Systems Development Manager Information Services Department
Pepsi-Cola Company, Purchase, New York, U.S.A

1. PROCEEDING

The successful use of business computer graphics typically develops in a major company from the company's growing use of large-scale computer systems. At Pepsi-Cola, and its parent, PepsiCo, the use of computer graphics evolved from the Company's desire to develop more efficient and effective methods of communicating information to decision-making executives, to identify opportunities and risks, and to find new tools to help with the decision-making process. The resulting graphics were made an effective business tool through systematically planning and executing presentation and interactive graphics. The determined effort at PepsiCo has brought the corporation to the forefront in the field of business computer graphics. Magazine/Newspaper articles in the United States have saluted PepsiCo's success in using computer graphics as a business tool and in integrating them into the corporation's operations.

Business computer graphics fall into two general categories at Pepsi-Cola: (1) Presentation Graphics, which illustrate various aspects of the company's business and (2) Interactive Graphics, in which company personnel are trained to create computer graphics for their part of the company's operations.

In the four years since the corporation began using computer graphics, the presentation graphic packages have concentrated on the areas of financial reporting, personnel administration, marketing and sales. Currently, there are two financial reporting packages, a personnel administration package, and three packages that show divisional marketing and sales volume performance.

All of these packages have been successful for the company, and each shares with the others common factors of success. First, each package provides global integration of information ("the complete business picture"). Second, each package is custom designed, supplying the information needed to monitor, control, and position the sector of the company using a particular package. Third, each package is supported with summary and detail reporting, with charts highlighting situations and identifying trends. Finally, the graphics packages are automatically updated and integrated with the company's data bases to incorporate the most accurate and current information available. Slides and copies are prepared upon request, giving an executive the ability to produce a tailored presentation.

Within Pepsi-Cola's Computer Graphic Facility, four operational hardware items are used to support presentation graphics: a color terminal, a color copier, a camera, and a digitizer. Four major types of charts are created, and samples of these charts are shown in the Graphic Portfolio, a sampler book used within the company. The major types of charts -- line, bar, pie and combinations -- plus maps and text are used individually and in combination to effect a wide variety of computer graphics.

The second main function of business computer graphics within Pepsi-Cola is interactive graphics. The personnel of the Computer Graphic Facility train users from throughout the company to create their own graphics and provide graphic equipment and consulting services. Training sessions address questions about types of output and concentrate on good chart design as well as selecting the right chart for a particular application. Additionally, the training includes the "whys and hows" of standardizing graphic formats and colors. To support this function, a easy-to-use user-friendly software package is utilized in conjunction with color CRTs and printers.

2. THE LARGE CORPORATION AND PRESENTATION COMPUTER GRAPHICS

As a major corporation, PepsiCo uses computer graphics both for overall corporate operations and for the operations of its primary business segments. The five business segments of PepsiCo are: (1) Beverages -- Pepsi-Cola Company and PepsiCo International; (2) Food Products -- Frito Lay and PepsiCo Foods International; (3) Food Service -- Pizza Hut, Taco Bell and PepsiCo Food Service International (4) Transportation -- North American Van Lines and LeeWay; and (5) Sporting Goods -- Wilson Sporting Goods. Although each of the five business segments functions in a decentralized fashion with its own plans and goals, each also operates in harmony with PepsiCo's plans and goals.

There are Information Services Departments at the World Headquarters for PepsiCo and its various business segments. For presentation graphics, the Information Services Departments provides the complete service for charts: analysis, design, and programming. In addition, it performs the production function of producing slides and copies.

Staff personnel in operating divisions are thereby able to concentrate on the business and their ability to do so is enhanced by the graphics developed by Information Services.

3. PRESENTATION GRAPHICS PACKAGES

The corporate Information Services Department utilizes two primary presentation graphics packages. The first focuses on financial reporting, while the second is a highly customized personnel system. The financial package is produced in hard-copy format for PepsiCo's vice president of finance and for the corporation's business planners. The package consists of 600 monthly charts that cover the corporation's operating divisions. In addition, approximately 100 to 150 slides are selected each month for review of the operating divisions.

Both the financial reporting and personnel graphics packages are supported with detail and summary reporting. Also, both packages provide the end user with the capability of producing slides and copies upon request.

Overall, the two corporate packages give management executives at World Headquarters a detailed and personal view of the diversified and widely-spread operations of a multi-billion dollar company.

Pepsi-Cola Company, the largest unit in Beverages, has two major presentation graphics packages that have been developed by the division's Information Services Departments. One package, consisting of 1,000 charts, is distributed to the division's senior staff, including the president. This package includes financial results, market performance and administrative statistics; while another Division Performance package of 85 charts shows sales performance. These two are prepared on a monthly basis. Second, the Market Review package, prepared bi-monthly and consisting of 900 charts plus selected slides, includes data collected by independent marketing companies such as A.C. Nielsen. The Information Services Department also develops about 500 slides per year for ad hoc presentations.

4. FACTORS OF SUCCESS

An examination of the common success factors for the presentation computer graphics packages for Pepsi-Cola reveals key characterics that are applicable to any graphics operation. In broad terms, the factors of success are that the graphics packages (1) provide global integration of information, (2) are custom designed, (3) are supported through summary and detail reporting, (4) are automatically updated for accuracy and timeliness, and (5) provide slides and copies upon request.

(1) Global Integration of Information. This feature gives the user the complete business picture -- the key information items an executive needs in order to manage effectively. Most executives manage profiles with a standard but complete approach, so the general format used at Pepsi-Cola should be transferable to other types of businesses. For example, the chief financial officer (CFO) manages operating divisions so there is a set of charts for each financial operating division. For the marketing executive there are 27 basic charts for each of the Nielsen major markets. The charts are informative and integrated, encompassing such factors as price, market share, and frequency of feature advertisements. Together the marketing charts give the user a market picture -- the information needed to make a decision. The charts provide the information necessary to answer many day-to-day business questions.

Not only is the package integrated, but individual slides can carry a complete picture. For example, a single slide may show a region's actual sales (compared to competitors), sales variances from the previous year's sales, and actual sales compared to planned sales. This type of slide, extremely informative yet simple, can be produced for all geographic territories and all products.

(2) Custom Design. A custom-designed graphics package can be a critical ingredient of overall graphics success. Through custom designing the business needs of a person or an organization are determined. Such a package must include features that answer the questions of what is needed to monitor and manage the business, what are the key leverage points of the business, and what information is needed to make day-to-day decisions.

(3) Summary and Detail Reporting. Since charts highlight situations, identifying trends, opportunities, and problems, detail reporting is needed to investigate and to provide explanations. A sudden dip in actual sales against planned sales highlights a detail that requires investigation.

(4) Automatic Updating. When graphics are updated on a regular basis, users of graphics packages can depend on their accuracy and timeliness. Through their dependability graphics become more effective.

(5) Slides and Copies Upon Request. Since slides and copies can be provided as they are called for, users of the computer graphics packages know that they can make presentations quickly and effectively. This function can greatly aid an executiv's decision-making.

5. HARDWARE AND SOFTWARE FOR PRESENTATION GRAPHICS

An effective grouping of hardware for presentation graphics includes a color terminal, a digitizer, a color copier, and a 35mm camera.

At Pepsi-Cola's Information Services Department, the following specific pieces of equipment have been used with notable success:

(1) Chromatic Color Terminal. The Chromatics CG1599, and 8-color interactive graphics computer with a 19 inch (48-cm) CRT display screen, can operate as a stand-alone computer or as a remote terminal. As a stand-alone, the Chromatics can be used to a develop diagrams and schematics, b receive graphic images from the bit pad digitizer, and c store graphics on floppy disks for subsequent redrawing. As a remote terminal the Chromatics produces or modifies graphic files. It interfaces with IBM 3033. Graphic images produced can also be stored on floppy disks for later redrawing and printing.

(2) Digitizer. The bit pad digitizer has proved particularly effective and inexpensive for mapping. An 11 inch (28-cm) square drawing area, the bit pad is electronically wired to transfer images to the Chromatics. Maps, logos, or free-style visuals are placed on the bit pad and their outlines traced. The bit pad transmits the coordinates of each point or line to the Chromatics terminal. Using Chromatics color-fill routines, the visual can then be colored and title and legends appended.

(3) Color Copier. The Xerox 6500, in addition to its functions as a color copier, has a control unit that enables it to receive the screen images directly from the Chromatics memory. Any graphic image displayed on the screen can be printed in full color on paper or overhead transparencies in single or multilpe copies.

(4) 35mm Camera. The Matrix 4007 camera is electronically wired to the Chromatics CG1599 so slides, instant prints, and overhead transparencies can be produced. A less expensive approach is to use a camera on a tripod in front of the Chromatics screen.

The software package Tell-A-Graf, was developed by Issco for the production of business charts. In order to create a chart, Tell-A-Graf requires a file of English-type instructions specifying the chart format and selection options such as the title, legend, and data to be plotted. These files may be created on a computer terminal under the CMS environment and can be accessed on the Chromatics terminal.

6. GRAPHICS OUTPUTS AND DISPLAY MEDIA TYPES OF GRAPHICS

The four major types of charts are line, bar, and pie charts, plus combinations of these. In addition, maps and text can be produced.

1. Line Charts. Line charts are particularly useful when illustrating data over a long period of time such as several years. Shading between lines can help emphasize comparisons. Computer graphics line charts can be enhanced by such techniques as putting a colored grid behind the lines to help identify the height of each line. Highlighting the zero line permits the quick communications of positive and negative information.

Symbols on the zero line can be used to pinpoint a specific time in each year.

2. Bar Charts. The two main types of bar charts are clustered and stacked, each of which can be shown vertically or horizontally. Clustered bar charts are best for showing the basis of different values next to each other; text annotation can highlight specific data. These are particularly popular when used horizontally as people are accustomed to reading left-to-right. With overlay effects such as a zero-based line, essential information is immediately visible to the user. Stacked bar charts, in which the different bar sections are placed on top of each other to create a cumulative value, can be confusing. The user must determine mentally the differences between bar sections, which can lead to inaccuracies and detracts from the graph's effectiveness.

3. Pie Charts. Pies, which can be annotated and/or broken into slices, are particularly effective at illustrating parts of the whole, as in sales contribution by brand. The slicing technique aids visual communication by emphasizing differences in slice sizes.

4. Combination Charts. Any chart types can be combined; the objective is to permit the user to see data in two or more perspectives. For example, a single slide can have a line chart showing sales for different brands over several years along side a pie chart illustrating percentage of total sales for each brand. Or, two bar charts can be put side by side when the user wants to see the relation of two similar entities -- such as total industry consumption of different categories of soft drinks and the sales in each category of a particular company.

5. Mapping. Mapping is especially important and useful in monitoring and planning marketing geographically. Mapping can be as large-scale as a country divided regionally or as specific as a city subdivided by section. At Pepsi-Cola, maps are created for computer graphics by tracing the map on the bit pad and then storing the outline of the maps on a disk. For each application the outline is copied, and titles and colors are added. The actual tracing is tedious, and problems can occur if connections are not made correctly when tracing. Then, colors spill through the gaps in the boundary of their designated areas on a map and spread into the adjacent area. However, the successful results of mapping are graphically dramatic and extremely useful.

7. INTERACTIVE GRAPHICS IN A LARGE CORPORATION

Interactive graphics, training a user to create graphics for his or her specific area of responsibility, can be very functional and effective in a large corporation. Although the users are then free to develop their own graphics and are therefore more disposed to recognize their effectiveness, Pepsi-Cola's Information Services Department offers essential support services by holding training sessions and workshops, by providing equipment and on-going consulting, and by insuring the quality of the graphics.

8. HARDWARE AND SOFTWARE

For Pepsi-Cola's interactive graphics program, two types of hardware are currently in operation: IBM 3279 color terminals and an IBM 3287 color printers. The software package is called Tag-Prompter. The 3279 CRT eight-color display terminals are under CMS on the IBM 3033. The terminals are connected to an IBM 3287 dot matrix printer, which receives data from the terminal and prints in red, green, blue, and black.

The Tag-Prompter software package, a front end to Tell-A-Graf, was developed to produce business charts easily. Tag-Prompter encourages the user to answer questions in order to create a chart. The package is very English-like and easy to use, qualities that have led to an enthusiastic reception by the users.

9. PLANNING AND PROMOTING INTERACTIVE GRAPHICS

Planning begins with setting an objective of user-created quality graphics. Some potential users will not know how to use a CRT, so a half-day terminal course can prove essential. Also, the development of training materials helps Information Services Department personnel to understand and master the products completely.

In order to promote computer graphics within a large corporation, it is necessary to make as many people as possible aware of their existence and of the ease with which they can be integrated into diverse business operations. By obtaining corporation organizational charts and sending an invitation -- with samples attached -- to a graphics overview, all potentially interested personnel can be reached. The overview presentation, with its dramatic color slides, can show visually the advantages of graphics over tabular reports and can convince potential users that graphics can be easy to create and use.

For those presentation attendees who express a desire to be trained, Pepsi-Cola offers a day-and-a-half session that includes classroom instruction, text materials, and a hands-on workshop. There is on-going consulting during the session with trained personnel at the equipment.

This type of promotion for interactive graphics has been very effective for Pepsi-Cola: 50% of the people trained have returned to the Information Services Department to create their own graphics.

10. TRAINING USERS IN INTERACTIVE GRAPHICS

Effective training of new graphics users can make interactive graphics as successful as presentation graphics. Good training of new users, coupled with on-going support and service, gives the new users confidence in creating computer graphics.

At Pepsi-Cola, training concentrates on how to choose and develop the right chart for the application. The course is divided into four sections: (1) Introduction, which expands on the purpose of the course; the types of outputs, which are slides, overhead transparencies, and paper; the types of charts, which are line, bar, and pie; and available graphics equipment, which has been described under Hardware and Software; (2) Chart Planning, which teaches good chart design and standards; (3) Chart Production, which instructs in the actual Tag-Prompter procedures; (4) Workshop, with hands-on experience.

Preliminary training in types of charts and the variety of available hardware and software is relatively straightforward. How to create and use charts is the most critical and difficult part of the training because most beginning users have never before worked with computer graphics.

11. CHART PLANNING

Chart Planning is probably the most important part of the training course, as selecting the correct chart is directly related to the effectiveness of the graphic. The user must be able to analyze the data in order to determine the chart's objective. If the user develops a focus sentence for the key information, the actual chart selected will be decided by the relationship of information. For example, training emphasizes that: (1) Pies show the relationship of each element to the whole; (2) Lines show the relationship of data over time, illustrating trends; (3) Bars show differences in performance for the same period and also show value quantitatively.

In Pepsi-Cola's training course, a typical problem in which the user should select a pie chart is as follows: a corporation has five store locations one of which (Las Vegas) outsells the others. The user creates the focus sentence, "Las Vegas sells more than any other branch store." Since Las Vegas is a part of the whole of branch stores, a pie chart is developed.

A bar chart problem also begins with a focus sentence: "All regions are performing better than the growth plan." A bar chart permits each region's sales growth to be measured against the planned growth for that region -- showing differences in performance.

12. CHART DESIGN AND PRODUCTION

After basic chart selection and planning, training should emphasize design. In a large corporation, there are several advantages to stressing standardization. With standard formats data readability is promoted and viewers can concentrate on the information presented rather than the format details. Further, when a standard format is seen repeatedly, a trend's visual pattern becomes familiar and unusual changes are recognized quickly.

Another advantage of standardization is that charts produced on a regular basis can be modified easily. After its creation the standard format is stored in the computer, from which it can be recalled each time the chart needs to be updated.

Standardization can be used for the size of the chart and its components -- placement of title, legend, text annotation, axis annotation, and grid and in color use. Color has meaning and aids in communicating messages. Color produces impact and permits selective emphasis. Color effectiveness can be enhanced by using company logo colors to identify company performance in all graphics. Or, planned objectives can always be the same color and actual performance another color.

The final stage of training is chart production, which at Pepsi-Cola shows the trainees how to use the language of Tag-Prompter. Workshop problems follow, with the user helped by a standard form to construct a chart. The form has spaces for the focus sentence, chart data, chart mock-up, and other information. Once a user is trained, he or she may use the facilities of the Information Services Department or may have equipment ordered.

13. CONCLUSION

Business computer graphics can meet successfully and dramatically the needs of large, diversified corporations for visual information. Through this relatively new business tool, executives and managers at many different levels of responsibility can make informed decisions more effectively and quickly. Presentation and interactive graphics used together can result in a successful approach to business computer graphics.

14. BIOGRAPHY

Mrs. Ulicny has been with Pepsi-Cola Information Services for the past eleven years. She currently holds the title of Manager, Information Center and is responsible for the development of information applications as well as micro computers and computer graphics. Mrs. Ulicny holds a B.A. in Psychology from West Virginia University and has done graduate work in Business Administration at Pace University. Additionally, she has participated in a Digital Computer education series as well as being appointed a reviewer of the National Computer Conference Association.

GPSR Compliance

The European Union's (EU) General Product Safety Regulation (GPSR) is a set of rules that requires consumer products to be safe and our obligations to ensure this.

If you have any concerns about our products, you can contact us on ProductSafety@springernature.com

In case Publisher is established outside the EU, the EU authorized representative is:

Springer Nature Customer Service Center GmbH
Europaplatz 3
69115 Heidelberg, Germany

Batch number: 10692442

Printed by Printforce, the Netherlands